VOLUME I

Computer Chips and Paper Clips

Technology and Women's Employment

Heidi I. Hartmann, Robert E. Kraut, and Louise A. Tilly, Editors

Panel on Technology and Women's Employment
Committee on Women's Employment and Related Social Issues
Commission on Behavioral and Social Sciences and Education
National Research Council

NATIONAL ACADEMY PRESS
Washington, D.C. 1986

National Academy Press • 2101 Constitution Ave., NW • Washington, DC 20418

NOTICE: The project that is the subject of this report was approved by the Governing Board of the National Research Council, whose members are drawn from the councils of the National Academy of Sciences, the National Academy of Engineering, and the Institute of Medicine. The members of the committee responsible for the report were chosen for their special competences and with regard for appropriate balance.

This report has been reviewed by a group other than the authors according to procedures approved by a Report Review Committee consisting of members of the National Academy of Sciences, the National Academy of Engineering, and the Institute of Medicine.

The National Research Council was established by the National Academy of Sciences in 1916 to associate the broad community of science and technology with the Academy's purposes of furthering knowledge and advising the federal government. The Council operates in accordance with general policies determined by the Academy under the authority of its congressional charter of 1863, which establishes the Academy as a private, nonprofit, self-governing membership corporation. The Council has become the principal operating agency of both the National Academy of Sciences and the National Academy of Engineering in the conduct of their services to the government, the public, and the scientific and engineering communities. It is administered jointly by both Academies and the Institute of Medicine. The National Academy of Engineering and the Institute of Medicine were established in 1964 and 1970, respectively, under the charter of the National Academy of Sciences.

This project has been supported by funding from the Women's Bureau of the U.S. Department of Labor, the National Commission for Employment Policy, the Economic Development Administration of the U.S. Department of Commerce, and by the National Research Council (NRC) Fund. The NRC Fund is a pool of private, discretionary, nonfederal funds that is used to support a program of Academy-initiated studies of national issues in which science and technology figure significantly. The NRC Fund consists of contributions from a consortium of private foundations including the Carnegie Corporation of New York, the Charles E. Culpeper Foundation, the William and Flora Hewlett Foundation, the John D. and Catherine T. MacArthur Foundation, the Andrew W. Mellon Foundation, the Rockefeller Foundation, and the Alfred P. Sloan Foundation; the Academy Industry Program, which seeks annual contributions from companies that are concerned with the health of U.S. science and technology and with public policy issues with technology content; and the National Academy of Sciences and the National Academy of Engineering endowments.

Library of Congress Cataloging in Publication Data

National Research Council (U.S.). Committee on Women's Employment and Related Social Issues. Panel on Technology and Women's Employment.
Computer chips and paper clips.

Bibliography: p.
Includes index.
Contents: v. 1. [without special title]
1. Women white collar workers—Effect of technological innovations on. 2. Office practice—Automation. 3. Microelectronics—Social aspects. 4. Women—Employment. 5. Women—Employment—Government policy—United States. I. Hartmann, Heidi I. II. Kraut, Robert E. III. Tilly, Louise A. IV:. Title.
HD6331.18.M39N38   1986   331.4'8165137'0973   86-18113
ISBN 0-309-03688-7

Printed in the United States of America

Panel on Technology and Women's Employment

LOUISE A. TILLY (*Chair*), Committee on Historical Studies, Graduate Faculty, New School for Social Research
TAMAR D. BERMANN, Work Research Institutes, Oslo, Norway
FRANCINE D. BLAU, Department of Economics and Institute of Labor and Industrial Relations, University of Illinois
DENNIS CHAMOT, Professional Employees Department, AFL-CIO, Washington, D.C.
MARTIN L. ERNST, Arthur D. Little, Inc., Cambridge, Mass.
ROSLYN FELDBERG, Massachusetts Nurses Association, Boston, Mass.
WILLIAM N. HUBBARD, JR., Hickory Corners, Mich.
GLORIA T. JOHNSON, International Union of Electronic, Technical, Salaried, and Machine Workers, AFL-CIO, Washington, D.C.
ROBERT E. KRAUT, Bell Communications Research, Inc., Morristown, N.J.
SHIRLEY M. MALCOM, American Association for the Advancement of Science, Washington, D.C.
MICHAEL J. PIORE, Department of Economics, Massachusetts Institute of Technology
FREDERICK A. ROESCH, Citicorp, New York
TERESA A. SULLIVAN, Population Research Center, University of Texas
DONALD J. TREIMAN, Department of Sociology, University of California, Los Angeles
ROBERT K. YIN, COSMOS Corporation, Washington, D.C.
PATRICIA ZAVELLA, Merrill College, University of California, Santa Cruz

HEIDI I. HARTMANN, Study Director
LUCILE A. DIGIROLAMO, Staff Associate
WILLIAM A. VAUGHAN, JR., Staff Assistant

Committee on Women's Employment and Related Social Issues

ALICE S. ILCHMAN (*Chair*), President, Sarah Lawrence College
CECILIA P. BURCIAGA, Office of the Dean and Vice Provost, Stanford University
CYNTHIA FUCHS EPSTEIN, Graduate Center, City University of New York, and Russell Sage Foundation, New York
LAWRENCE M. KAHN, Department of Economics and Institute of Labor and Industrial Relations, University of Illinois
GENE E. KOFKE, Montclair, N.J.
ROBERT E. KRAUT, Bell Communications Research, Inc., Morristown, N.J.
JEAN BAKER MILLER, Stone Center, Wellesley College
ELEANOR HOLMES NORTON, Georgetown University Law Center
GARY ORFIELD, Department of Political Science, University of Chicago
NAOMI R. QUINN, Department of Anthropology, Duke University
ISABEL V. SAWHILL, The Urban Institute, Washington, D.C.
ROBERT M. SOLOW, Department of Economics, Massachusetts Institute of Technology
LOUISE A. TILLY, Committee on Historical Studies, Graduate Faculty, New School for Social Research
DONALD J. TREIMAN, Department of Sociology, University of California, Los Angeles

Contents

CONTENTS, VOLUME II viii
PREFACE xi
ACKNOWLEDGMENTS xv

1. TECHNOLOGICAL CHANGE AND WOMEN
   WORKERS IN THE OFFICE 1
   Technological Change, 6
      Information Technologies, 7
      Social Context of Technological Change, 10
      Output and Employment: Trends and Interpretations, 13
   Women's Employment, 18
      Overview, 18
      Why Technology May Affect Women Differentially, 19

2. HISTORICAL PATTERNS OF TECHNOLOGICAL
   CHANGE 24
   The Telephone and Telephone Operators, 25
   Workers in Printing and Publishing, 29
   The Automated Office and Its Workers, 32
      Secretaries, 33
      Accountants and Bookkeepers, 38
      Insurance Clerks, 40
      Bank Tellers, 44
   Retail Clerks, 48
   Nursing and Nurses, 52
   Conclusions, 58

3. EFFECTS OF TECHNOLOGICAL CHANGE: EMPLOYMENT LEVELS AND OCCUPATIONAL SHIFTS 62
   Problems in Employment Projections, 63
      Underlying Factors, 63
      Data Problems, 66
   The Supply of Women Workers, 68
      Labor Force Participation Rates, 68
      Projections of Labor Force Participation Rates, 70
      Projections of Age-Specific Rates, 72
      Other Features of Women's Labor Force Participation, 73
      Educational Attainment of the Labor Force, 75
   The Potential Effects of Technological Change, 79
      The Influence of Labor Supply, 79
      The Demand for Workers, 81
      Unemployment, 83
   Recent Trends in Clerical Employment, 86
      Overall Growth, 86
      Occupational Shifts Within Clerical Work, 88
      Demographic Trends in Clerical Employment, 89
      Sources of Change in Clerical Work, 96
   Outlook for Clerical Employment, 103
      Overall Growth, 103
      Occupational Shifts, 111
      Job Loss and Displaced Workers, 124
   Conclusion, 125

4. EFFECTS OF TECHNOLOGICAL CHANGE: THE QUALITY OF EMPLOYMENT 127
   Employment Quality, 127
      Defining Employment Quality, 129
      Workers' Satisfaction and Attitudes, 131
      Job Content: Job Fragmentation and the Deskilling Debate, 136
      Working Conditions, 143
      Economic Considerations, 148
      Conclusion, 150
   Implementing Technological Change and Improving Employment Quality, 150
      The Role of Managers, 151
      The Role of Workers, 157
      Conclusion, 165

5. CONCLUSIONS AND RECOMMENDATIONS 167
   Summary, 167
   Education, Training, and Retraining, 170
   Employment Security and Flexibility, 172
   Expansion of Women's Job Opportunities, 173
   Adaptive Job Transitions, 175
   Identification and Dissemination of Good Technological
      Design and Practice, 177
   Worker Participation, 178
   Monitoring Health Concerns, 179
   Data and Research Needs, 179
   Epilogue, 181

REFERENCES 183

BIOGRAPHICAL SKETCHES OF PANEL MEMBERS
AND STAFF 201

INDEX 207

Contents

Volume II: Case Studies and Policy Perspectives

I. OVERVIEW
   Technology, Women, and Work: Policy Perspectives
      Eli Ginzberg

II. CASE STUDIES OF WOMEN WORKERS AND INFORMATION TECHNOLOGY
   The Technological Transformation of White-Collar Work:
      A Case Study of the Insurance Industry
         Barbara Baran
   "Machines Instead of Clerks": Technology and the Feminization
      of Bookkeeping
         Sharon Hartman Strom
   New Technology and Office Tradition: The Not-So-Changing
      World of the Secretary
         Mary C. Murphree
   Integrated Circuits/Segregated Labor: Women in Three
      Computer-Related Occupations
         Myra Strober and Carolyn Arnold

III. TECHNOLOGY AND TRENDS IN EMPLOYMENT

Women's Employment and Technological Change: A Historical Perspective
Claudia Goldin

Recent Trends in Clerical Employment: The Impact of Technological Change
H. Allan Hunt and Timothy L. Hunt

Restructuring Work: Temporary, Part-time, and At-Home Employment
Eileen Appelbaum

IV. POLICY PERSPECTIVES

Employer Policies in the Application of Office System Technology to Clerical Work
Alan F. Westin

New Office and Business Technologies: The Structure of Education and (Re)Training Opportunities
Bryna Shore Fraser

The New Technology and the New Economy: Some Implications for Equal Employment Opportunity
Thierry J. Noyelle

Managing Technological Change: Responses of Government, Employers, and Trade Unions in Western Europe and Canada
Felicity Henwood and Sally Wyatt

Preface

Striking advances in microelectronic and telecommunications technology have transformed many worlds of work. These changes have revolutionized information storage, processing, and retrieval, with immediate and long-range consequences for clerical work. Since women—nearly 13 million of them—are the overwhelming majority of clerical workers, they are and will be disproportionately affected by this type of technological change. Jobs may be created or eliminated, but they have also been and will certainly continue to be transformed. So far, knowledge about these large processes of change has been scattered and incomplete. There is great need for more systematic evaluation and understanding of this technological change and its specific effects on the conditions of and opportunities for women's employment.

In light of this need, the Committee on Women's Employment and Related Social Issues established its Panel on Technology and Women's Employment in March 1984. The tasks of the panel included gathering together what is now known on the subject; identifying areas in which research is most needed and commissioning scholars to undertake research for the committee; preparing this report, which discusses the available research and proposes both research and policy recommendations; and organizing a conference to present the findings and recommendations. The panel's work was supported by the Women's Bureau of the U.S. Department of Labor, the National Commission for Employment Policy, the Economic Development Administration of the U.S. Department of Commerce, and the National Research Council Fund.

The panel benefited greatly from earlier work of National Research Council committees that have posed and examined questions about women's employment. *Women, Work, and Wages: Equal Pay for Jobs of Equal Value* (1981), the report of the Committee on Occupational Classification and Analysis, surveyed both earnings differentials between men and women and the relationship of these differentials to occupational segregation, and evaluated the usefulness of job evaluation plans and occupational reclassification as tools to achieve pay equity. The panel's parent body, the Committee on Women's Employment and Related Social Issues, established in 1981, has produced a volume of essays, *Sex Segregation in the Workplace: Trends, Explanations, Remedies* (1984), and a full report, *Women's Work, Men's Work: Sex Segregation on the Job* (1986).

This earlier work provides background material on the issues that faced the panel: the rapid introduction of new technologies in clerical work in the last five years and contradictory interpretations about its consequences in the short and longer run. To what extent do current changes differ from earlier ones? Is the new microelectronic and telecommunication technology creating or eliminating jobs? In what ways is it affecting the quality of employment for those whose job organization and content are being transformed? Are there differential effects that depend on the skill, occupation, industry, or demographic characteristics, such as minority status or age of workers? If jobs disappear or change drastically, what kind of support—training, retraining, relocation—might be needed for displaced workers? What institutional arrangements might be necessary or desirable for planning and implementing change or devising support programs?

The panel's answers to these questions are contained in this volume. Many of the research papers commissioned by the panel are published in *Volume II: Case Studies and Policy Perspectives*. (A list of the contents of Volume II precedes this preface.)

On the basis of its examination of available data and research that identifies the recent and possible future effects of technological change on the quantity and quality of women's paid employment opportunities, the panel expects that, over the next 10 years, the sometimes contradictory patterns of the present will continue. The changes—both the slowdown of growth in clerical employment and the ongoing shifts in the distribution of clerical jobs—merit policy attention because of their magnitude. Under these conditions, some workers are likely to be caught in unforeseen transitions and become unemployed. Minority women, older women, and those with low levels of education or training may find such transitions especially difficult. For new entrants to the labor market it may be hard to find entry-level jobs. Changes in job content may cause skill mismatch between available workers and available jobs. We do not expect that the current rate or type of technological change will be so great as to require fundamental alterations in employment policy regarding women. We offer policy recommendations directed toward easing what could be difficult transitions for some workers in terms of lost jobs or reductions in employment quality even in the

PREFACE

best case. The large degree of uncertainty about our central conclusion, however, leads us to propose more far-reaching policies in the event that the future holds a more severe set of changes than we now expect. Recommendations about research and data needs are also offered.

In sum, this report reveals the relationships that exist between and among various groups, of which women workers are but one, who share significant interests in solving problems linked to technological change, yet maintain differences about solutions and the distribution of costs. The panel's recommendations are designed quite explicitly to highlight those relationships and shared interests in order to promote agreement about goals and the means to achieve them.

LOUISE A. TILLY, *Chair*
Panel on Technology and
Women's Employment

Acknowledgments

A report such as this is a collective product, and it is a pleasure to thank the many people involved in producing it. The Panel on Technology and Women's Employment consists of academic scholars in several relevant disciplines, experts in the design and application of technology, and business and labor leaders. Although panel members held differing views and frequently voiced differences of opinion, each one contributed generously to the group endeavor. Drafts of specific parts of the report were prepared by both individuals and working groups, and many individuals prepared memoranda commenting on those early drafts. Panel member Robert Kraut deserves special acknowledgment for drafting all of Chapter 4. The process of integrating the materials and sections written by members, consultants, and the study director was truly collective. That process was exemplary; I appreciate it deeply and thank all those involved.

In carrying out its tasks, the panel commissioned a review of recent research findings, an inventory of data sources, an analysis of trends in clerical employment, and 14 scholarly papers. The panel also held a workshop at which researchers presented their findings in areas identified as central to the problem. These materials, as well as the valuable and informed intellectual exchange on fundamental issues between authors and panel members, were significant contributions to our inquiry.

In addition to the panel members and authors of commissioned papers, I would like to express my appreciation to the panel's staff. The study director, Heidi I. Hartmann, took major responsibility for overall rewriting and editing of the varied prose styles and sometimes intellectually untidy contributions of the panel. This she accomplished in addition to coordinating the process, recruiting expert papers, herself serving as an expert, finding data and analyzing them when needed, and administering the ongoing work of the Committee on Women's Employment and Related Social Issues. She has been a strong and incomparably valuable resource for the panel; we sincerely thank her for her important contribution to our efforts. We also thank Lucile DiGirolamo, staff associate, for her organizational expertise and unflappable calm. In addition to organizing the panel's meetings, recording minutes, and planning the dissemination conference for our report, she kept track of the numerous source materials used in our work. William A. Vaughan, Jr., served as staff assistant during the second year of the project, as did Katherine Autin and Rita Conroy during the first year. They aided our work in innumerable ways, not the least of which was the word processing of our report. Commission staff Diane Goldman, Christine McShane, Beverly Blakey, and Suzanne Donovan also contributed to the report. Micaela di Leonardo, now at Yale University, served as a consultant to the panel early in its work; she contacted many researchers working in the field and solicited their cooperation in our efforts. Jackie George of Wheaton College and Victoria Threllfall of Bennington College served as interns at early and late stages of the project, respectively, and provided research assistance. To all of them we owe our thanks.

Several members of the Committee on Women's Employment and Related Social Issues—Cynthia Epstein, Lawrence Kahn, and Isabel Sawhill—reviewed a draft version of this report; we thank them for their prompt and useful response. We also thank Alice S. Ilchman, chair of the committee, for the many ways in which she facilitated the panel's work. Members of the Commission on Behavioral and Social Sciences and Education and the monitor appointed by the Report Review Committee of the National Academy of Sciences thoroughly reviewed a draft of the report and made several helpful suggestions. In addition, the report was helpfully reviewed by several experts outside the Academy structure: Vary Coates of the Office of Technology Assessment of the U.S. Congress; Carol Romero, Sara Toye, and Stephen Baldwin of the National Commission for Employment Policy; and Allan Hunt and Timothy Hunt of the Upjohn Institute. Eugenia Grohman, associate director for reports for the commission, contributed substantially to the report's clarity through her insightful editing. David A. Goslin, executive director of the commission, has our appreciation for his continued support of the work of the committee and its panels.

Several organizations made this report possible through their financial support. We thank both the organizations and their representatives who provided liaison with the panel. At the Women's Bureau of the U.S. Department of Labor, we thank Collis Phillips, Mary Murphree, and Roberta McKay. In addition, we would like to recognize the interest and strong support of the project shown by Lenora Cole-Alexander, former director of the Women's Bureau. Carol Romero and her staff at the National Commission for Employment Policy

ACKNOWLEDGMENTS

provided much helpful information as well as sustained interest. Beverly Milkman and Richard Walton, at the Economic Development Administration, U.S. Department of Commerce, aided us with a grant to allow us to complete the project in a timely matter. Crucial early funding was provided by the National Research Council Fund.

LOUISE A. TILLY

Computer Chips
and Paper Clips

1
Technological Change and Women Workers in the Office

The employment effects of technological change have again become an issue of public concern. In particular, questions have been raised about future employment opportunities for women workers because many of the new developments—especially in telecommunications and microprocessing—have already dramatically altered work involving information processing, an area of work dominated by women. Recent research suggests that automation in the clerical sector is altering the numbers and types of jobs available, the nature of jobs and their geographic location, working conditions, and career opportunities. A major quantitative study recently predicted that the number of clerical jobs in the economy will decline by 1995, not only relatively but also absolutely (Leontief and Duchin, 1984), although this prediction is controversial. Earlier waves of automation in office work also raised fears of unemployment, yet clerical, professional, technical, and managerial employees have increased their share of the labor force in every decade since at least 1940 (Hunt and Hunt, 1985a).

Is the present situation different? For the past several decades, as employment in the U.S. economy has continued to shift from agriculture and manufacturing to services, the latter sector has absorbed large numbers of new workers, particularly women and young people. Technological developments have contributed to the introduction of new products and services that provide new kinds of jobs. Sectoral shifts in demand, accompanied by economic growth, historically have produced better opportunities and wages for some workers. Two striking examples for women workers are the nineteenth-century shift of young women workers from agriculture to manufacturing and the recent shift of black women from agriculture and domestic service to clerical positions. Economywide, technological change is credited with contributing to productivity growth, standard of living increases, and generally higher wages. In the last decade, service industries have adopted new strategies, reorganized work, and introduced new machines to increase productivity. If productivity gains increase substantially in the service sector, will it continue to generate enough jobs to absorb all the workers seeking employment? What kinds of jobs will they be? Will they offer safe and healthy working conditions, job stability, good wages, and opportunities for advancement? Both future levels of employment and the "quality" of work are closely related to the introduction and use of new technologies.

This report identifies and analyzes the effects of technological change on both the quantity and quality of women's paid employment. It focuses particularly on innovations in information processing and telecommunications and their applications in offices—past, present, and likely future. The interest of the Panel on Technology and Women's Employment is in determining whether women may be differentially affected by these innovations relative to men and, if so, how; to what extent women of different ages, educational backgrounds, and race and ethnic groups may be affected differently; and what factors may be shaping these effects.

The "machine" aspects of technological change in office automation have captured the public imagination, but they are only part of the picture. They have drawn the attention of scholars, workers, employers, and policy makers to a much more pervasive set of changes in the organization of work, its geographic location, and the characteristics of workers. Hence, when the report examines employment effects resulting from technological change in microprocessing and telecommunications (information technology or "telematics"), it includes the wider changes in work organization and composition of the labor force as well as the more immediate effects of the new machines themselves.

In seeking to determine the effects of technological change on women's paid employment, the report focuses on new technologies in clerical occupations, both because the technical developments in these occupations are dramatic and appear to have been widely implemented and because so many women work in these jobs. Clerical occupations are a diverse group, ranging from the "typical" office ones of secretaries, typists, and file clerks to bill collectors, interviewers, telephone operators, dispatchers, mail carriers, insurance adjusters, bank tellers, and proofreaders.[1] Table 1-1 displays the number of workers in clerical

---

[1] There are many definitions of clerical occupations; conceptually, clerical work is not a clear category, and the boundaries are difficult to delineate. This report generally uses the 1980 census classification of "administrative support workers" as its definition of clerical workers, but notes significant changes that have occurred in the government classifications. The most typical clerical occupations, those that would be regarded as clerical workers in any classification system—secretaries, stenographers, typists, file clerks, bookkeepers, accounting and financial clerks, and general office clerks—account for 8.9 million, or more than half of the 16.8 million workers identified as administrative support workers in the 1980 census.

occupations in the 1980 census (16.9 million), with the number and percent female in each. The five clerical occupations with the largest number of workers were secretaries (3.9 million), bookkeepers and accounting clerks (1.8 million), general office clerks (1.6 million), typists (0.7 million), and general office supervisors (0.6 million). For women clerical workers, the first four largest occupations were the same, but general office supervisors (0.4 million) were in seventh place, below receptionists (0.5 million) and bank tellers (0.5 million). (Cashiers, another large occupation—1.7 million and about 85 percent female—often thought of as clerical workers, were reclassified in the 1980 census as sales workers.) The 13.0 million women clerical workers represented more than three-fourths of all clerical workers and more than one-third of all employed women workers in 1980.

In reviewing data and research on employment trends and on the development and implementation of new information technologies in several clerical occupations and sectors, the report seeks to identify both the positive and negative aspects of this wave of technological change and to reduce the uncertainty surrounding estimates of the likely size, incidence, nature, and timing of the effects. The report also considers how employers' decisions to adopt and implement new technologies and workers' participation in these processes can affect outcomes. Finally, the panel identifies and recommends private and public policies that can alleviate negative effects, promote improvements in women's employment opportunities, and produce a more humane work environment for all.

The remainder of this chapter first defines technological change and describes some recent and anticipated changes in information technologies. It then discusses the measurement and characteristics of technological change and explores in a schematic way the economic and social context of technological change and the employment effects that can result. It next describes women's employment situations and explains why the panel expects differential effects from technological change for women workers.

Chapter 2 discusses selected examples of technological change in communications and information processing and considers especially its effects on women's employment levels, absolutely and relative to men, and on the content and quality of work. The effects discussed include indirect, unforeseen, and uneven effects. These topics are examined in several critical types of women's employment: communications work and clerical work in information and data processing, with a comparative look at retailing and nursing. The chapter also considers briefly how workers have responded to technological change in these cases and how managers' and workers' interests have differed and coincided.

Chapter 3 analyzes the current and future effects of technological change on the levels of employment and the structure of occupations and considers changes in both the supply and demand of workers. It reviews projections made by the Bureau of Labor Statistics and presents the panel's estimates of the most

TABLE 1-1 Employment of Administrative Support Occupations in 1980

Occupation	Total Employment	Percent Female	Female Employment
Administrative support occupations	16,851,398	77.1	12,997,076
Supervisors, administrative support occupations	1,056,710	47.1	497,668
Supervisors, general office	631,337	56.1	354,410
Supervisors, computer equipment operators	42,142	29.4	12,392
Supervisors, financial records processors	157,409	49.0	77,172
Chief communications operators	66,765	34.3	22,898
Supervisors, distribution, scheduling, and adjustment clerks	159,057	19.4	30,796
Computer equipment operators	408,475	59.0	241,155
Computer operators	384,392	58.9	226,354
Peripheral equipment operators	24,083	61.5	14,801
Secretaries, stenographers, and typists	4,656,955	98.3	4,579,938
Secretaries	3,870,582	98.8	3,823,248
Stenographers	85,785	90.7	77,841
Typists	700,588	96.9	678,849
Information clerks	894,178	85.4	763,561
Interviewers	134,002	78.0	104,582
Hotel clerks	61,217	68.2	41,756
Transportation ticket and reservation agents	99,449	57.5	57,161
Receptionists	516,498	95.8	494,800
Information clerks, n.e.c.	83,012	78.6	65,262
Nonfinancial records processing	965,107	77.2	745,372
Classified-ad clerks	13,552	77.6	10,521
Correspondence clerks	19,309	81.5	15,741
Order clerks	311,321	67.4	209,871
Personnel clerks	75,235	87.4	65,759
Library clerks	140,731	81.2	114,294
File clerks	277,592	79.7	221,350
Records clerks	127,367	84.7	107,836
Financial records processing	2,254,084	88.4	1,991,619
Bookkeepers and accounting clerks	1,827,890	89.7	1,640,233
Payroll clerks	159,292	83.3	132,622
Billing clerks	129,380	88.9	115,020
Cost and rate clerks	85,855	68.4	58,731
Billing, posting, calculating machine operators	51,667	87.1	45,013
Duplicating, mail, office machine operators	58,671	65.6	38,462
Duplicating machine operators	18,822	61.0	11,484
Mail and paper handling machine operators	7,052	62.3	4,390
Office machine operators, n.e.c.	32,797	68.9	22,588
Communications equipment operators	308,690	89.5	276,148
Telephone operators	292,165	91.0	265,938
Telegraphers	7,604	35.7	2,711
Communications equipment operators, n.e.c.	8,921	84.1	7,499

TABLE 1-1 (*Continued*)

Occupation	Total Employment	Percent Female	Female Employment
Mail and message distribution clerks	773,826	29.6	229,096
Postal clerks	267,035	35.8	95,511
Mail carriers, postal service	256,593	12.9	33,179
Other mail clerks	167,973	47.3	79,425
Messengers	82,225	25.5	20,981
Material recording, scheduling, and distributing	1,662,256	34.4	571,300
Dispatchers	94,830	31.2	29,568
Production coordinators	254,625	44.2	112,539
Traffic, shipping, and receiving clerks	481,958	23.6	113,554
Stock and inventory clerks	570,906	34.7	198,345
Meter readers	41,407	10.2	4,239
Weighers, measurers, and checkers	72,040	36.6	26,348
Samplers	2,542	45.5	1,157
Expediters	106,146	53.9	57,242
Material recording, n.e.c.	37,802	74.9	28,308
Adjusters and investigators	515,666	62.3	321,234
Insurance adjusters, examiners, investigators	163,586	60.2	98,407
Noninsurance investigators and examiners	243,616	62.4	151,951
Eligibility clerks, social welfare	24,128	81.8	19,744
Bill and account collectors	84,336	60.6	51,132
Miscellaneous administrative support occupations	3,296,780	83.2	2,741,523
General office clerks	1,648,934	82.1	1,353,251
Bank tellers	494,851	91.2	451,465
Proofreaders	27,321	79.1	21,610
Data-entry keyers	378,094	92.4	349,477
Statistical clerks	139,174	75.0	104,345
Teachers' aides	206,695	92.7	191,564
Administrative support, n.e.c.	401,711	67.2	269,811

NOTE: n.e.c., not elsewhere classified.

SOURCE: Data on total employment from Hunt and Hunt (1985a:Table 2.1(a)); data on female employment and percent female from Hunt and Hunt (1985a:Table 2.4); based on 1980 decennial census data.

plausible worst case. The chapter also examines recent and projected shifts in the demand for various occupations.

Chapter 4 analyzes the current and likely future effects of technological change on employment quality and examines how job content changes in terms of autonomy, responsibility, and knowledge. The chapter considers workers' attitudes and job satisfaction, computer-based monitoring and pacing, telecommuting, and other aspects of job quality. It also explores the roles of managers and workers in implementing new technologies and improving employment quality.

Chapter 5 identifies some remaining areas of uncertainty, particularly with regard to the panel's analysis of likely future change. The panel offers recommendations to facilitate the mutual adaptation of technology and employment. The panel concludes that the problems posed by the new technologies (particularly for women workers) are capable of solution (1) through public and private policies that increase opportunities for women workers to benefit from change—by means of education, training, and equal employment opportunity—and that provide assistance to those who become unemployed as a result of change; (2) through the development of models for collaborative decision making and their dissemination; and (3) through research on the effects of technology and alternative ways to implement it.

TECHNOLOGICAL CHANGE

Much recent technological change involves the systematic use of new scientific knowledge in the production of goods and services. Earlier technological change, which could be called inventive technology, was episodic and unpredictable; its experiential, noncumulative character limited its applications. Starting around the turn of this century, however, scientific research began to substitute for random inventions and to produce solutions to socially defined problems, and methodical scientific research now feeds technology with potential solutions on a continuous, regular basis. This relatively recent approach has become the dominant mechanism of technological change, although discontinuous invention by imaginative individuals has by no means disappeared: for example, that quintessential twentieth-century invention, the Xerox photocopy method, was developed in 1938 by a creative patent attorney working independently with an unemployed physicist; they could not sell their idea to any large office machine supplier (*Washington Post,* August 21, 1985:B1-B2).

Built on a scientific base in addition to inventors' efforts, applications have become more flexible; contemporary technological change both provides more choice and generates continuing change. Today, technological change is something humans decide to do, and it can be directed in ways to produce positive results. Scientific knowledge and inventive technology make change feasible; social decisions, shaped by cultural attitudes as well as economic considerations, determine where change occurs and what is produced.

Technological change alters the goods and services that are produced or the methods by which they are produced. It is presumed that such alterations increase productivity or competitive advantage. These alterations produce increased output with the same inputs (or reduce or make less costly the inputs needed to produce the same output), improve the quality of the outputs, or create entirely new products or services. Technological change is often conceived simply as the application of innovation in science and engineering to production—for example, the development of rubber, the telephone, penicillin, ceramics, or computers. However, it also encompasses new organization of production or distribution processes—the assembly line, typing pool, or self-service. Innovation and the reorganization of production also occur amidst general changes in the uses of capital and labor and the location of both these factors. Technological change and its concomitants, by this very broad definition, have the potential to affect all areas of social life, including art and literature. This report confines its attention to technological change that occurs in the workplace and has an impact on paid employment, particularly on women workers in clerical jobs. Other non-workplace-based technological changes—for example, birth control and household technologies—have had enormous effects on women's paid employment in this century, but they are not examined here.

INFORMATION TECHNOLOGIES

The most important technological developments affecting women's work are those involved in the miniaturization of computers and the dramatic improvements in telecommunications. Both developments affect the speed and ease of manipulating, analyzing, and communicating information, tasks in which the vast majority of women workers are employed. With the advent of the microprocessor, small computers with as much memory as the early large mainframe computers (which in their earliest vacuum-tube design filled entire rooms) became feasible. Sufficient circuits for their central processing units, including work space and operating system, can be placed on one silicon chip as small as one-quarter square inch. With the development of sophisticated software that requires little expertise in computer languages, small computers have become ubiquitous. They are incorporated in home appliances, automobiles, and hand-held calculators, as well as in stand-alone word processors and home and personal computers. The miniaturization of processing is proceeding apace. The number of components that can be placed on a silicon chip has doubled every 12 to 18 months for the past 20 years. Experts see the near-exponential growth in chip capacity continuing; denser chips perform increasingly complex logical tasks faster. The cost of chips has decreased a thousandfold in 20 years (Ross, 1985). Advances in telecommunications have also occurred rapidly. If developments in aircraft design had occurred on the same scale as those made in chips, "planes carrying 500,000 passengers each would be flying between New York and London for a fare of 25 cents" (Ross, 1985:35).

Computers

Some observers of the very rapid increases in processing speed and memory size of computers suggest that the cost of computing hardware will continue to fall. They also expect a very rapid diffusion of computers even to small businesses, so that by 1990 perhaps one-half of all offices will have some form of computer available for office tasks. They point to annual increases in sales of 13 to 16 percent and to increasing capital/labor ratios in much of the service sector: in financial services, for example, capitalization is at $8,000 to $10,000 per worker, compared with approximately $25,000 to $30,000 in manufacturing, but the gap is closing (Office of Technology Assessment, 1985). Other observers point out hidden costs (such as software development, software maintenance, system maintenance) and remaining technical difficulties (such as establishing local area networks that link personal computers to each other and to a mainframe) and suggest that diffusion will be much slower. These observers also point to very recent decreases in sales of new computer equipment and consequent layoffs among vendors, which they take as an indication of sources of resistance to computerization in the office (Hunt and Hunt, 1985a). Still other observers point to established social arrangements in offices that may make the full adoption of office automation less feasible (Murphree, 1985; Iacono and Kling, 1986; see also Chapter 2 of this report). Finally, the rate of adoption will of course depend on the capacities and prices of the equipment offered, which will be affected by the rate of technical change within the computer-manufacturing industry and in telecommunications. Recent developments in these areas are considered next.

Data-Entry Technologies A keyboard is currently the dominant mode of data entry, but other methods are being developed. For example, an optical character reader (OCR) is available in many workplaces for specific uses and selected type styles. The OCR currently available is 40 to 50 times faster than keyboard entry. Although OCR capacities are increasing quite rapidly, it will probably be some time before handwriting can be interpreted reliably by machine. Speech recognition and voice-activated "typewriters" are another potential input methodology, but technical development here is proceeding more slowly; most machines are limited to selected speakers and small vocabularies. Nevertheless, some experts expect that by the year 2000, 35 percent of data entry will be by OCR or speech recognition.

Storage and Processing For microcomputers, magnetic discs are the dominant form of storage, and their storage capacities have increased several times since their first appearance. Optical discs, currently being developed, can store up to 25 times more than magnetic discs at one-thirtieth the cost; 250,000 pages can be stored on one optical disc (Office of Technology Assessment, 1985). Improvements are also being made in image processing; the development of software (because of increasing reliance on already-developed subroutines); the creation of expert systems, such as medical diagnostic systems, which are intended to simulate experts in analyzing information; and electronic mail, including software for "automatic calendaring"—the process of keeping track of appointments and arranging mutually convenient meetings.

Output and Display Technologies Rapid change is occurring in improving display capabilities: better-quality printers, such as the laser printer, the liquid crystal screen, and more flexible and powerful plotters, all of which will become available in multiple colors. Speech synthesis, a nonvisual output, is also developing rapidly. It is already used in several large-scale applications, such as in telephone number information services.

System Compatibility and Interconnection System interaction is clearly crucial if services such as electronic mail are to become widely used and if microcomputer users are to have access to a variety of information. It appears, however, to be a very difficult area. Although a report by the Office of Technology Assessment (1985) indicates that the use of local area networks (LANs) is expanding rapidly, some businesses are currently having trouble connecting personal computers and mainframes within their individual organizations; substantial programming is usually involved, which slows the process. This problem could be at least partially solved by standardization, which is not now occurring. In the absence of significant standardization, peripheral conversion equipment is being developed.

Telecommunications

Telecommunications has also improved dramatically in the past two decades and is increasingly integral to computing capability. The recent report of the Office of Technology Assessment (1985) suggests that computers and telephones are becoming increasingly alike. The interrelation between communications and computing is easily seen: for example, a single large (but standard) switch in the Atlantic Telephone & Telegraph system has 1.6 million lines of programming (Ross, 1985). The major components of telephony are a customer terminal, transmission, switching, and signaling.

Over the years new functions and capabilities have been added to the telephone handset, the most common customer terminal today; other types of terminals in use include teletypewriters, data terminals, and computer terminals. Each of these terminals can incorporate microprocessors, giving them "logic" and "memory," making them "intelligent" and capable of more functions. The handset now provides call waiting, call forwarding, three-way calling, and so on, and sometimes serves as a terminal for data entry and information retrieval (for example, at-home banking). An earlier technical advance was the ability to transmit the digital information used by computers over the analog lines of the telephone system with the use of the modem (named for its function, modulator-demodulator). Increasingly, telephone services are installing digital lines, which will speed transmission of data and video material as well as voice.

The technical problems in transmission involve distance and volume. Transmission today uses copper wire, coaxial cable (concentric copper cables), radio (including microwave and satellite), or optical fiber cable. A very large coaxial cable can carry 132,000 simultaneous conversations. Optical fiber is small and lightweight, with very high capacity (approximately 1,400 simultaneous conversations on a single fiber), relatively immune to electrical interference and resistant to signal loss, and may eventually replace all other cable. A transoceanic optical fiber cable is being developed for use in the late 1980s. Optical fiber cables transmit information in digital form, which is more flexible than analog form, since voice, video, or data can be transmitted in digital form.

Telephone switching has undergone enormous technical change. Today's automatic switching devices handle more than 600,000 calls per hour. Signaling takes place over a parallel network that provides the means of signaling and controlling transmission, switching, and customer terminals. The signaling system allows long-distance calls to be switched completely from their origin to their destination in less than two seconds.

Among the developments that are foreseen over the next decade in computing and telecommunications are improvements in the rapidity of computer response, greater responsiveness to the user in the simultaneous accommodation of several computer languages, increased use of telecommunications to connect computers and to distribute data and computation where needed, development of local area networks, increased ease of human communication with computers through voice recognition and more natural computer languages, and sophisticated and highly portable software (Bucy, 1985). All of these changes have affected and will continue to affect information-processing jobs by decreasing the cost of that processing and increasing its ease. Coupled with increased transportation capabilities, these innovations also increase the flexibility with which office work can be performed. For example, the increased miniaturization and decreased cost of computers makes work at home on small computers feasible. Large "back offices" in remote locations can serve the national needs of a company with retail branches or executive offices located elsewhere. And, within offices, innovations have contributed, and will continue to contribute, to reorganizing work and changing its nature.

Social Context of Technological Change

Technological change, narrowly defined as the progressive adoption of a particular innovation, generally proceeds unevenly; an innovation takes time to diffuse throughout the economy until its use becomes the new norm. Historically, the time required from invention to first commercial application of an innovation has usually ranged from 5 to 20 years, and the period from first use to widespread use, from 1 to 15 years; diffusions of major innovations, such as electricity, have taken up to 50 years (Katz and Lazarsfeld, 1955; Mansfield, 1966). Electronic computing has been used in offices since the late 1950s, when large mainframe computers first became available. Operated by specialized data-processing staffs and custom programmed for the particular applications needed, they were first used to assist in large-scale and repetitive tasks: payroll, statistical analysis of large data sets like the national census, and so on. Interactive systems, like that for airline reservations, were developed in the 1960s. Now microcomputers with direct telephone hookup promise a new office form: the integrated workstation, where one person can quickly and simultaneously have access to information from several data bases.

Although technological change may be readily conceptualized as the adoption of particular innovations, it is in fact very difficult to measure technological change embodied in innovations and to assign specific results to particular innovations. Two examples—the assembly line and word-processing equipment—suggest some of the measurement problems. The assembly line, a major technological innovation, contributed to reducing labor input per automobile and to lower sales prices for automobiles. In the same period, personal incomes increased, movies spread new tastes and behavior, and suburbanization created housing spatially removed from central cities. The automobile became a mass-consumption product—leading to a substantial increase in output and employment in automobile manufacturing. The decline in labor input per automobile resulted in an aggregate employment gain, but how much of the gain can be directly attributed to the assembly line?

Word-processing equipment may well contribute to increases in the ease and speed of word entry and to reductions in the cost per page entered, but it may also contribute to increased or more consequential error because of the automated capacities, for example, a deleted page or a "lost" document (one that was not properly stored electronically). Furthermore, as many people have observed, word processors may lead to greater numbers of drafts of a given manuscript and so to more time spent in word processing. Again, an innovation that might be expected to save labor and lead to reductions in employment might actually increase employment. And, as word processors give way to personal computers and workstations, more far-reaching effects may occur: the lines between secretarial and professional/managerial work may blur as secretaries increase their access and ability to manipulate information, allowing them to generate analytic reports; the location of work may become more flexible with stand-alone equipment and telephone access to centralized data bases; and a reduced need for paper record-keeping may alter work organization dramatically.

These examples illustrate the ways in which the uses and effects of technology are interrelated with many other factors, including the social organization of work, the location of housing, and the norms of a community. Innovations do not by themselves increase productivity or alter employment levels; the ways they are implemented, coupled with changes in other factors, cause them to have these or other effects. The context in which new technology is introduced influences its effects on productivity, the quantity of employment, job quality, and the work environment.

One consequence of the twentieth-century pattern of science-based technological change has been an increase in choices about alternative paths and applications by decision makers, although the degree of choice is not always apparent. For any given technology, there is usually a range of choices in design and implementation. Equipment may constrain, but it does not determine, work organization or job quality. The substantive scientific base of current technological developments in telematics increases flexibility between technical capabilities and job outcomes. Microprocessors can be programmed and used in a variety of ways: for example, software can be more or less customized, more or less interactive. There can be considerable freedom in how managers and workers use information-processing equipment, but the degree of choice they exercise will depend on their knowledge and authority in the organization. These factors vary among workers and managers as well as between them. Although the same piece of equipment can be used in different ways to achieve any given goal, goals are not always clearly articulated and, like knowledge and authority, also differ among the parties involved. The political and social organization of the firm as well as its economic situation will thus affect the choices made about the implementation of technology.

Although it is socially embedded, technology is at the same time and in itself a source of dynamism in society simply because it is likely to alter the continuing feasibility of change and shape its direction. It is difficult to predict the uses of new technologies: some promising ones fall by the wayside; others lie fallow for many years. The majority of innovations introduced to the market fail, but others far surpass anything imagined when they were initially developed: the use of the phonograph and radio for entertainment was not envisioned by their inventors; when computers were first developed it was thought that 8 to 10 computers would be sufficient to meet the needs of the entire United States because only a few organizations were thought to be of sufficient size to use them efficiently. The unpredictability of the long-term consequences of technological change has itself become commonplace in our history and our expectations. Having experienced the changes brought about by industrialization, assembly lines, automobiles, airplanes, telephones, televisions, charge cards, computers, gene splicing, and nuclear reactors, people in a modern society no longer imagine life without substantial technological changes—even if no one knows what they will be.

Output and Employment: Trends and Interpretations

In the economy as a whole, both output and employment have grown substantially since World War II. Nonfarm employment grew nearly 70 percent between 1948 and 1978, while output increased nearly 165 percent (Mark, 1979). However, employment, output, and productivity growth have slowed since 1974. Past experience has given rise to several contradictory interpretations of this slowdown. One is that "too much" technological change has caused the decline of employment growth (Sadler, 1981; Noble, 1984). Advocates of this interpretation note that a long-term tendency for unemployment to increase seems to be developing, despite productivity improvement in the recent recovery. They fear that economic growth based on rapid technological change may not encourage employment growth—that the widespread use of microcomputers may contribute to the employment problem, not solve it. The nature of the new computer-based technologies has also led to questions about whether the educationally disadvantaged will become even more relatively disadvantaged if they do not have appropriate training.

A second interpretation holds that greater investments in technology are necessary to enhance productivity and to increase output, to keep the United States internationally competitive, and to maintain a healthy economy in which employment will grow (Schmitt, 1983; Adler, 1984b). Indeed, several U.S. industries, such as steel, textiles, and shoe manufacturing, that failed to innovate and restructure have become less and less internationally competitive. Exponents of this interpretation point to the increase in the number of persons employed, including many women, in such industries as banking and insurance, which have been technologically innovative industries. Although the rate of growth has slowed in some occupations and a few have nearly disappeared, service sector employment has grown overall because demand for its products and services has burgeoned. New microprocessing and telecommunication technology have both reduced costs of old services and permitted the introduction of new products and services in banking, insurance, and other industries (Appelbaum, 1984); the innovations have thus contributed to increases in demand. Because these changes also involve new forms of organization, recruitment, and training, some of these observers predict that open opportunity and access to training may lead to a more democratic workplace rather than to further disadvantaging the disadvantaged.

A third, and the most futuristic, interpretation of the effect of technological change on employment suggests that in the foreseeable future, productivity may be so enhanced that employment may become a rarity for everyone. People will need to learn how to use vastly increased leisure time well; new social mechanisms, other than employment, will have to be developed to distribute income (Leontief, 1983; Bell, 1984). As these conflicting interpretations illustrate, new technology poses uncertainties, problems, and possibilities.

Output Measures

Most measures of technological change are economywide and indirect. The most commonly used indicator of technological change is the increase in output per hour of labor input, or *labor productivity*. What has been the change in gross national product (GNP) relative to the increase in hours worked? The labor productivity measure attributes all changes in output, including those caused by increased input of capital, to increases in the productivity of labor. A more accurate indicator of technological change, *total productivity,* relates output to total inputs, or at least to both capital and labor. But measurement and interpretation problems persist with both labor input and total input measures: for example, both methods measure changes in *quantity* better than changes in quality. Furthermore, measures of the output of all services, but particularly of government services, are based almost totally on inputs (see National Research Council, 1979).

Total productivity measures, which estimate the average rate of technological change at approximately 2 percent per year, attribute to technological change all output growth other than that due to changes in the quantity of the inputs. It is estimated that as much as 90 percent of the growth in output per capita since 1900 has been due to technological change and macroeconomic factors other than changes in the quantity of capital and labor (Mansfield, 1966; Hunt and Hunt, 1985b). Foremost among these other factors associated with greater output is improvement in the quality of labor, particularly through increases in average educational attainment.

Other, somewhat more direct measures of technological change attempt to measure scientific and technical effort: for example, changes in the amount of money spent on research and development, in the number of patents issued, and in the number and proportion of scientific and technical personnel. Case studies of specific innovations in selected industries have also been carried out. They are generally able to assess how rapidly a particular innovation was adopted throughout an industry or several industries and how effective it was in improving productivity. Such direct measures, however, are difficult to interpret independently of effects in the economy as a whole.

Employment Effects

Identifying and measuring the employment effects of technological change are even more difficult than measuring the output effects of technological change. The employment effects may not occur until long after the introduction of an innovation; they may occur in entirely different firms, industries, or locations or entirely different occupations from those actually using the innovation; or they may be entirely unanticipated. Furthermore, as with output changes, employment effects are qualitative as well as quantitative, and qualitative change is always much more difficult to measure. Finally, as noted, it is difficult to isolate the effects of technology from the effects of other economic and social changes occurring simultaneously with the introduction of new technologies; this inseparability is the fundamental problem in the interpretation of employment effects.

Employment Levels In the production of a particular product or service, it is axiomatic that if technological change increases the productivity of labor while demand for output does not increase, employment must fall—if not immediately, eventually; if not in the particular firm that innovates, in another that does not; if not in the number of workers, in the number of hours worked; if not in laying-off present workers, in failing to hire new workers. But productivity growth can also increase demand for output by lowering its price, increasing its quality, or creating new products that tap new demands, so that employment can even increase substantially.

History provides examples of each of these effects. In U.S. agriculture, rapid technological change linked with modest growth in demand for output led to substantial displacement of workers from the agricultural sector. In 1900 nearly 12 million workers, or 40 percent of the labor force, were employed in agriculture; today, agricultural employment is just over 3 million, or 3 percent of the labor force. In automobile manufacturing, rapid technological change led to substantial reduction of automobile prices and to increased demand and greater employment in the early to mid-twentieth century. More recently, in telephone communications, high productivity growth from rapid technological change between 1960 and 1975 was coupled with substantial increases in output and employment; the number of people employed declined in some occupations, such as telephone operator, while it increased in others, such as customer service representative.

Overall technological change and shifts in consumer demand have resulted in an employment shift to services over the last several decades. In 1950 about 48 percent of the civilian employed labor force (59 million workers) worked in the goods-producing sector (agriculture, mining, construction, and manufacturing), while 52 percent worked in the service-producing sector (transportation and public utilities; wholesale and retail trade; finance, insurance, and real estate; business and personal services; and government). By 1980 only 31 percent of the labor force (99 million workers) worked in the goods-producing sector, while 69 percent worked in the service-producing sector (*Monthly Labor Review,* January 1985; calculated from Tables 1 and 9). The percentage increase in employment in the service-producing sector was 143 percent, compared with 39 percent in the goods-producing sector. Among the industries within the service sector with the fastest-growing employment over the 30-year period were government (169 percent); finance, insurance, and real estate (173 percent); and business and personal services (234 percent).

Technological change coupled with shifts in consumer demand for specific products is, of course, only one source of change in the level of employment. Changes in aggregate demand are equally if not more important. Unusual changes in labor supply also contribute to changes in employment and unemployment. The baby-boom cohorts born after World War II were associated with a dramatic increase in young entrants to the labor market that exacerbated youth unemployment during the 1970s and 1980s. These decades will be followed by several in which the relatively small cohorts of the 1970s and 1980s will enter the labor market.

According to a recent study in the *Monthly Labor Review* (Podgursky, 1984), unemployment has increased substantially over the past several business cycles. Between 1969 and 1982, unemployment at the troughs increased by 4.8 percentage points, and unemployment at the peaks increased by 3.8 percentage points. One explanation for this long-term increase in unemployment is the rapid growth in the labor force, especially among youth and women; unemployment from this source could be viewed as frictional, associated with the normal difficulties new entrants have in finding work. Podgursky finds this source an important factor in the early 1970s, but by the late 1970s and early 1980s, prime-age men were the group whose unemployment contributed most to the increase. This finding is consistent with his finding that blue-collar workers in manufacturing contributed disproportionately to the unemployment increase over the period. This result in turn suggests that structural shifts in demand were important, although Podgursky argues that the data suggest a possible overall slackening in aggregate demand as well.

Concerns about levels of employment were the major motivation for an important earlier examination of the relation between employment and technological change. Twenty years ago the U.S. National Commission on Technology, Automation, and Economic Progress (1966) issued its report and six volumes of supporting studies. With the general increase in employment that has occurred since the late 1960s, the questions that motivated it are relevant again today: "Had the pace of technological change accelerated until the economy could no longer make adequate adjustments? Was technological change a major cause of persistently high general levels of unemployment? Was the new technology so twisting the demand for labor that the undereducated and unskilled were becoming unemployable while the demands for highly trained manpower were insatiable?" (Bowen and Mangum, 1966:1). The commission found that fears of massive unemployment because of technological change were unwarranted: the pace of technological change had not accelerated much if at all, and even the skill mismatch was exaggerated. However, the commission did predict that the situation for minority workers and for youth would worsen substantially if appropriate training programs were not instituted, a prediction that has proved to be correct. Despite the fact that women's rapid entrance into the labor force had already begun, it is interesting to note that not a word was said about any special needs or different experiences that women might have in the labor market.

Employment Quality The commission's report was also relatively silent on another aspect of employment—the quality of work. Today, concern about changes in employment quality is voiced along with concern about quantity. At a 1982 international conference on office work and the new technology, organized by the Working Women Education Fund, speaker after speaker warned of the dangers to job quality posed by the new office technologies (Marschall and Gregory, 1983). Machung (1983), for example, warned of secretarial work becoming deskilled, repetitive, and monotonous as it is transformed into word-processing work. In its 1983 report, *The Future of Work,* the AFL-CIO Committee on the Evolution of Work warned that automation would lead to downgrading of many jobs in both factories and offices.

Some features of new technology suggest that it could contribute to a reduction in the quality of the work environment in some jobs. Early experience with the application of microprocessing and telecommunications in the clerical sector have led to reports of discomfort or possibly more serious physical problems that accompanied the use of video display terminals. Some workers perceived their autonomy and opportunity for career development to have been reduced. The new technologies have the capability of monitoring the worker's output and behavior more closely, possibly increasing stress in the job (Feldberg and Glenn, 1983). For example, in 1983 and again in 1984, an information operator was threatened with dismissal for exceeding the C&P Telephone Company's 30-second average work time per call by 3 seconds; the operator insisted she was morally obligated to assist customers in need of extra help and to provide a higher standard of service (*CWA News,* March 1983, May 1984).

But the new technologies also have the potential to improve work quality as they reduce drudgery and promote more integrated work processes. Many workers have welcomed the challenge of learning new skills and mastering complex systems. The important point is that whatever the reason for adopting new equipment and organization of work, that very adoption opens choices that have consequences for the quality as well as the quantity of employment. The choices of technology made by designers, producers, purchasers, and implementers (whether these seek profit, competitive advantage, a higher-quality product, or other goals) and the way that they organize the work process in relation to equipment may produce positive or negative effects, or both, on the quality of employment.

WOMEN'S EMPLOYMENT

OVERVIEW

Women's employment has risen rapidly since World War II and especially in the past 25 years. The increase in women workers accounted for 60 percent of the growth in the labor force in the past decade and is expected to account for 70 percent of the growth in the next decade. The range of occupations in which women worked also grew, as occupations held by both men and women grew faster in the last decade than the traditionally female occupations, and women entered many predominantly male occupations formerly closed to them. Table 1-2 shows changes in the distribution of women workers across occupations for the past three decades.

Women's wages still remain low relative to men's, however. The ratio of women's to men's wages for full-time, year-round work has averaged around 60 percent for several decades. In 1983 women who worked full time, year round averaged $13,468, 64.8 percent of men's average of $20,000 (Women's Bureau, U.S. Department of Labor, 1985). Some researchers predict that this

TABLE 1-2 Major Occupation Groups of Employed Women, 1950–1980 (percent)

Occupation	1950	1960	1970	1980
Total women	100.0	100.0	100.0	100.0
White-collar workers	52.5	56.3	61.3	63.5
Professional	12.2	13.3	15.5	15.9
Managers	4.3	3.8	3.6	6.8
Clerical	27.4	30.9	34.8	33.8
Sales	8.6	8.3	7.4	7.0
Blue-collar workers	43.9	41.8	37.9	35.5
Crafts	1.5	1.3	1.8	1.8
Operatives	20.0	17.2	14.8	10.7
Laborers	0.9	0.6	1.0	1.3
Private household	8.9	8.4	3.9	3.0
Other services	12.6	14.4	16.3	18.8
Farm workers	3.7	1.9	0.8	1.0
Managers	0.7	0.6	0.2	0.3
Laborers	2.9	1.3	0.6	0.7

SOURCE: Bianchi and Spain (1984:Table 3).

gap will close as women workers' years of experience in the labor force come to approach more closely those of men [(Smith and Ward, 1984; Goldin, 1985); see, however, Treiman and Terrell (1975) and Corcoran and Duncan (1979), who show a strong relationship between gender and earnings even when years of labor force experience are similar for women and men]. Others believe that reductions in discrimination will bring about a smaller wage gap (Blau and Ferber, 1986). In general, the percent female of an occupation is strongly correlated with its average earnings: the more women in an occupation, the less it pays. At present, job segregation by sex in the labor market is still substantial and continues to affect women's earnings and career mobility. And significant evidence suggests that promotional opportunities and access to on-the-job training for women, which are at least partially determined by employer actions, are restricted relative to those for men (Reskin and Hartmann, 1986).

Women continue to devote more of their time off the job to home, family, and child care than do men (Hartmann, 1981), and these family responsibilities, which are especially burdensome for minority women, may constrain their educational and labor market opportunities. Hispanic women tend to have larger families, and black women are more likely than white women to be single parents with the total responsibility for raising children and financially supporting them.

Because women workers tend to be concentrated in a limited set of occupations, because they are sometimes less geographically mobile, and because their access to education, training, or promotion within and across firms may be more limited than men's, the panel expects technological change to affect women and men differently. Because women earn less than men, they may also have fewer resources with which to respond to technological change. If technological change affects women's employment opportunities more negatively, or less positively, than it affects men's, it will contribute to maintaining, or even worsening women's relatively disadvantaged status. Conversely, technological change, if managed toward that goal, could provide a means to equalize the status of women and men.

WHY TECHNOLOGY MAY AFFECT WOMEN DIFFERENTIALLY

Job Segregation

Job segregation by sex has been large and relatively stable in the United States, although the past two decades have seen some decline. Women work largely in different occupations than do men, and they are in occupations that are predominantly female. More than 36 percent of all employed women work in just 10 occupations, and 9 of these are female dominated: secretaries, elementary school teachers, bookkeepers, cashiers, office clerks, "managers—

not elsewhere classified," waitresses and waiters, salesworkers, registered nurses, and nursing aides. Only 1 of the 10 largest occupations for women workers, "managers—not elsewhere classified," is among the 10 largest occupations for men (see Table 1-3). Job segregation by sex decreased in the past decade both because female dominated occupations grew less rapidly than in the past and because a substantial number of occupations became more integrated, particularly in the professions and management. The clerical occupations, however, for the most part became more female dominated (Reskin and Hartmann, 1986).

TABLE 1-3 Employment in the 10 Largest Occupations for Men and Women, 1980

Ten Largest Occupations for Men Detailed 1980 Census Occupational Title and Code	Number of Men	Percentage Female		1970–1980 Change in Percentage Female
		1980	1970	
1. Managers, n.e.c. (019)	3,824,609	26.9	15.3	11.6
2. Truckdrivers, heavy (804)	1,852,443	2.3	1.5	0.8
3. Janitors and cleaners (453)	1,631,534	23.4	13.1	10.3
4. Supervisors, production (633)	1,605,489	15.0	9.9	5.1
5. Carpenters (567)	1,275,666	1.6	1.1	0.5
6. Supervisor, sales (243)	1,137,045	28.2	17.0	11.2
7. Laborers (889)	1,128,789	19.4	16.5	2.9
8. Sales representatives (259)	1,070,206	14.9	7.0	7.9
9. Farmers (473)	1,032,759	9.8	4.7	5.1
10. Auto mechanics (505)	948,358	1.3	1.4	−0.1

Ten Largest Occupations for Women Detailed 1980 Census Occupational Title and Code	Number of Women	Percentage Female		1970–1980 Change in Percentage Female
		1980	1970	
1. Secretaries (313)	3,949,973	98.8	97.8	1.0
2. Teachers, elementary school (156)	1,749,547	75.4	83.9	−8.5
3. Bookkeepers (337)	1,700,843	89.7	80.9	8.8
4. Cashiers (276)	1,565,502	83.5	84.2	−0.7
5. Office clerks (379)	1,425,083	82.1	75.3	6.8
6. Managers, n.e.c. (019)	1,407,898	26.9	15.3	11.6
7. Waitresses and waiters (435)	1,325,928	88.0	90.8	−2.8
8. Salesworkers (274)	1,234,929	72.7	70.4	2.3
9. Registered nurses (095)	1,232,544	95.9	97.3	−1.4
10. Nursing aides (447)	1,209,757	87.8	87.0	0.8

NOTE: n.e.c., not elsewhere classified.

SOURCE: Rytina and Bianchi (1984).

Clerical Occupations

The growth in clerical employment during the post–World War II period undoubtedly facilitated the increase in female labor force participation that occurred over the same time period. Between 1940 and 1980, the proportion of all workers employed as clerical workers doubled from about 10 to 20 percent, while female participation in the labor force nearly doubled. The proportion of women working as clerical workers also increased substantially. Clerical occupations remain the mainstay of women's employment, and it is these occupations that are currently undergoing substantial technological and organizational changes integral to broader structural changes in the service sector. The service sector disproportionately employs clerical workers and women. Changes in either the level of employment or the quality of work in that sector will directly, and disproportionately, affect women. In 1981 only 6.3 percent of men worked in clerical occupations, compared with 34.7 percent of women.

Clerical occupations are diverse along several dimensions. Some are better paid and require more skill than others. Some are held disproportionately by minority women; others disproportionately by majority women. Earlier waves of automation were expected to decrease the number of clerical workers, but the number continued to increase. Earlier automation not only had displacement effects in individual occupations but also substantial employment-increasing effects across occupations. Recently, innovations in computation, dictation, and record-keeping have affected various occupations in different ways: since 1950, the number of stenographers has declined dramatically; the number of typists, file clerks, and postal clerks has decreased since 1970; and the number of computer operators increased substantially between 1950 and 1980 (Hunt and Hunt, 1985a).

Scholars disagree about how to interpret emerging trends and to project the future effects of technology on levels of clerical employment. In their well-known study, Leontief and Duchin (1984) offer disaggregated forecasts for the next 20 years which suggest that, relative to other occupations, clerical work will be more affected by displacement due to computer-based automation. In contrast, the Bureau of Labor Statistics (Silvestri et al., 1983; Silvestri and Lukasiewicz, 1985) predicts substantial growth of clerical workers through 1995. The contradictory forecasts stem from differences in assumptions about the rate of innovation and diffusion of various new technologies, their productivity effects, and the size of changes in final demand over time for the products and services clerical workers help to produce. Whichever prediction turns out to be more accurate, however, the concentration of women in clerical occupations and the concentration of technological change in the same occupations suggest that women are likely to be differentially affected relative to men. Of course, if women had equal opportunity in the labor market, they would almost certainly be less concentrated in the clerical occupations and technological change there would have less differential effects on women.

Differential Status and Access to on-the-Job Training

Even when they are in the same occupations, men and women may face different opportunities because they work in different firms with different pay levels (Blau, 1977; Strober and Arnold, 1985). Even within the same firms, women and men are likely to have different access to on-the-job training and promotion (Duncan and Hoffman, 1979). Employers' attitudes toward workers and their treatment of them are often conditioned by gender (Reskin and Hartmann, 1986). These stratification patterns may signal potentially negative effects of technological change for women in general; in addition, minority women experience ethnic and racial prejudice as well as sex-based discrimination, which constrains their opportunities further (Malveaux, 1982). For example, minority women may be placed in back-office jobs in which access to promotion and training opportunities is limited. Women may also be differentially affected relative to men because as workers, men and women do not enjoy the same status and power or access to organizational and collective resources (such as labor unions).

Differential Responsibility for Family Care

Even if men and women shared equal opportunity to benefit (or suffer) from workplace technologies, women might be prevented from taking advantage of opportunities because of their greater responsibility for family care and housework. They may also be more geographically restricted, less able to participate in educational programs, and more constrained in their job choice. One aspect of the capability of the new technologies may be especially relevant to differences between women and men. The microprocessor may make home-based production economically feasible for employers, and women's household responsibilities and the lack of affordable child care may make homework especially attractive to women. Part-time homework, either on a salaried or self-employed basis, may be preferred by some women. The new technologies may also lead, however, to increases in involuntary part-time and temporary work, if they have significant employment-displacing effects.

Because women have traditionally provided more care for family members than men have and because men—and many women too—have seen this arrangement as right and proper, some managers have developed assumptions about women workers that are increasingly inaccurate today. One such historic assumption is that women workers have higher turnover rates and less commitment to their jobs than men because of their family obligations (Feldberg and Glenn, 1979). This assumption may lead to less concern about eliminating women's jobs or to reorganizing their work in potentially inequitable ways, such as in the creation of temporary or part-time work with limited opportunities. But changing technology and reorganization can also be an opportunity for opening up new occupations and increasing mobility for women if the changing characteristics of women workers, in particular their increasing attachment to the labor force, are taken into account.

Conclusion

The panel expects that differential effects of the new technologies with respect to sex are likely. And for some of the same reasons—differences in choices, opportunities, treatment—we expect that some effects will also differ among subgroups based on ethnic, racial, age, marital, educational, or geographic characteristics of women. In many areas relevant to the new technologies, however, the effects are essentially the same for all workers and all subgroups of women. The new technologies have monitoring capabilities, for example, whether *either* women or men work with them, or *both* men and women do. The work environment is equally important for all workers. This report therefore discusses issues that apply to all workers, although they are particularly relevant for women. It focuses primarily on technological change in clerical occupations.

2
Historical Patterns of Technological Change

Contemporary observers of the recent innovations in microprocessing and telecommunications technology and their diffusion often claim uniqueness for the rapidity and nature of those changes. Historians point out, however, that the characteristics of change can only be determined after change has played itself out, if such a moment can be isolated; that the rapidity or completeness of change can only be judged in comparison with previous changes; and that any assumption that technology itself is the cause of change can only be verified by historical investigation of the context of change. As C. Wright Mills (1956: 193) pointed out more than 30 years ago with reference to his generation's office machines: "Machines did not impel the development, but rather the development demanded machines, many of which were actually developed especially for tasks already created."

Processes of change belong to history in two fundamental ways. First, they take time to unfold; anyone who looks only at a moment of the process—including the present moment—runs a great risk of mistaking its character. Second, they cling to time and place; how they happen varies significantly from one time and place to others, as a direct consequence of events in previous times and places. People who want to understand these large processes must examine them in their historical contexts (Tilly and Tilly, 1985).

This chapter uses history as a guide for understanding the complex relationship of technological change and women's employment. The chapter discusses both the historical characteristics of the relationship and the manner in which technological change, both in the more distant and the recent past, has been linked to changing levels of women's employment and to the quality of their work. Five cases of the effects of innovation on women's occupations are examined here. Three of the five cases involve communications and information-processing occupations that have been shaped over the long run by electronic technologies. The first case takes an innovation—the telephone—as its focus, examines its development and diffusion, and traces the history of the telephone operator, the women's occupation most tightly linked to that invention. The second case concerns a manufacturing industry—printing and publishing—and its workers in two waves of technological change. The third case looks at two sets of information-processing clerical occupations—word processors such as secretaries, and data processors such as accountants, bookkeepers, insurance clerks, and bank tellers—over the long period from the introduction of the first mechanized devices in the nineteenth century to the more recent introduction of electronic word and numeric data processing. For purposes of comparison with regard to the strength or weakness of the ability of women workers to shape change affecting them, the fourth example concerns retailing and its clerks, and the fifth, nursing, one of the quintessential women's professions.

All the cases examine three potential sources of change in employment levels: (1) economic growth, both overall and within industries (which results in more jobs); (2) loss of jobs because of technological innovation; and (3) substitution of women for men workers, or vice versa, in an industry. The cases further discuss, to the extent possible, the relationship of technology and the quality of work in affected occupations and industries. Finally, response to change, by both unionized and nonunionized workers, is also examined in situations of technological change. Throughout this examination it should be remembered that, historically, changes in levels of employment and quality of work have been complex, caused by many factors, and contingent—shaped by impersonal forces as well as by actors.

THE TELEPHONE AND TELEPHONE OPERATORS

The telephone is of interest both because it is the basis of an industry that has been and still is a major employer of women and because of its distinctive characteristics as an invention. De Sola Pool (1977a:3–4) describes the telephone as

a facilitating rather than a constraining device . . . [it] seem[s] to defy definition of even the primary effects; these seem polymorphous though indubitably large . . . the study of the telephone's social impact belongs to the important and subtle class of problems in the social sciences which demands a logic more complex than that of simple causality—logic that allows for purposive behavior as an element in the analysis.

Early in the history of the telephone, while its inventor and certain entrepreneurs visualized a fantastic future, other businessmen and engineers failed to understand the uses to which the telephone would be put and the extent to which it would be used. Even today there is scholarly disagreement over what indirect effects the telephone has had on people's social lives.

Bell had his historic conversation with Watson on March 10, 1876. By March 1878, Bell had envisioned a national switched network of business and residential telephones for interpersonal communication, through which one telephone subscriber could talk to any other in the country, even though this vision was technologically infeasible at the time (de Sola Pool, 1977b:156). In contrast to Bell's semiprescience, others of the time were less sure of the telephone's potential. Shortly after Bell's perfection of the telephone, for example, William Orton, the president of the Western Union Telegraph Company, turned down an offer to buy all rights to its patent with the words, "What use could this company make of an electronic toy?" (Aronson, 1977:16). Soon after, Sir William Preece, the chief engineer of the British post office, testified to a special committee of the House of Commons that the telephone had little future in Britain (de Sola Pool, 1977b:128):

> I fancy the descriptions we get of its use in America are a little exaggerated, though there are conditions in America which necessitate the use of such instruments more than here. Here we have a superabundance of messengers, errand boys and things of that kind.... Few have worked at the telephone much more than I have. I have one in my office, but more for show. If I want to send a message—I use a sounder or employ a boy to take it.

Social commentators trying to understand the social impact of the telephone had an even less successful record than did Bell and other entrepreneurs predicting its technological and business impact, partly because the latter were themselves in some position to shape the future, while the former could only imagine it. The commentators, in fact, often focused on the wrong issues, for example, emphasizing the role of the telephone in promoting world peace or in encouraging or reducing crime. They failed to recognize that the telephone was a facilitating technology.

The compression of space that the telephone allows had diverse effects (Gottmann, 1977). The telephone allowed companies to move away from their suppliers and customers and to concentrate in urban areas; it permitted management to establish corporate headquarters separate from their factories. It also substituted conversation over wires for the slowness of within-city mail, allowing companies to do business more economically through electrical communication within and between large office buildings. Even as it fostered urban congestion through business concentration in cities, the telephone, like the automobile and earlier transportation improvements, also promoted the suburbanization of housing.

Early telephone service was expensive because of technical limits; in 1896 telephone service cost $20 per month in New York. Charging for message units was the solution that opened the service to small consumers. However, as more distant links were added, complexity and cost continued to increase until fully automatic switching was achieved. This improvement made possible more local calls per worker at lower cost, starting in the late 1920s. Great reductions in the time necessary to complete long-distance connections, in the number of operators needed to put a call through, and in cost came after World War II. Toll calls per operator rose from 70 per hour in 1950 to 20,000 per hour in 1980 (Kohl, 1986).

The employment effects of the telephone have not always proceeded as predicted or, indeed, as intended. The first commercial telephone operators were boys, following the pattern of the older telegraph industry. Women quickly replaced the boys, who apparently were unruly tricksters when faced with the opportunity for tinkering with wire connections and teasing customers. According to a contemporary, "the work of successful telephone operating demanded just that particular dexterity, patience and forebearance possessed by the average woman in a degree superior to that of the opposite sex" (cited in Maddox, 1977:266). The operators' jobs, clean and dignified, were also attractive to educated, middle-class women. The Bureau of the Census in 1902 added an economic explanation for the employment of women: "Telephony, with its simpler, narrow range of work to be performed at the central office, has provided opportunity for a large number of young girls at a low rate of pay, comparing in this respect with the factory system" (quoted in Baker, 1964:69). The spread of telephone use that accompanied economic growth opened up this new occupation to women across the nation.

Being an operator was a stressful job, however, combining a good deal of physical effort, supervisors' and customers' demands for precision, and constant interpersonal contact. Early studies noted a high incidence of nervous illness and enormous turnover. In response, the Bell system became a pioneer in offering fringe benefits like vacations and sick pay (Maddox, 1977:269–270). Nevertheless, unionization began early. With the help of the Women's Trade Union League, operators in Boston organized in 1908 and led a strike of New England operators against Bell in 1919.

Although the physical demands of the job have decreased, pressure on operators to handle calls rapidly has not; the form of the pressure has been automated as electronic monitoring of operator efficiency has taken the place of supervisory surveillance. The telephone operator job continues to be one with high turnover, "a classic dead-end job," as Laws (1976) called it. In this case, technology was adopted in ways to promote efficiency with little attention to protecting or improving job quality.

What was the effect of the technology on levels of employment? Obviously, in its overall effect, the telephone stimulated job growth in cities. For operators, the years of the depression coincided with the first introduction of automatic dialing and the subsequent reduction of jobs. A study of one city's switch to a dial system in 1930 showed little dismissal of permanent operators: the company planned the changeover several years in advance and took advantage of the turnover rate (40 percent per year) to hire new operators on a strictly temporary basis for the two years preceding the change (in Baker, 1964:240–241). A less sanguine overview (Anderson and Davidson, 1940:426) found that "in view of lower living costs, the 32 percent fewer workers who were still employed in 1933 were slightly better off than they had been formerly. For the third of the workers eliminated from the service and thrown on a flooded labor market, however, the situation has become dire." They argued, however, that although automation in telephone switching would reduce the labor force, increases in service would lead to continued increase of supervisory and business staff. Nevertheless, Anderson and Davidson (1940:428) concluded: "Technology has its way, obviously, only when it will reduce operating costs and yield larger profits. This usually means fewer workers in proportion to volume of business and a proportionately lower wage bill." In the long run, this view proved correct with regard to telephone operators.

World War II saw an enormous increase in communications needs, including both increased demand for operators and improved services. By 1950 there were 342,000 telephone operators. The number declined subsequently to about 250,000 in 1960 and 184,000 in 1964. By 1978 there were 166,000 operators for 98 million telephones; in 1900 there had been 100,000 for 7 million telephones (Scott, 1982). New jobs for women appeared in the telephone company business offices because of the increased number of customers and services offered (Baker, 1964:245, 246). Because the administrative side of the telephone industry hired many women, the loss of women operators on the "production" side was compensated. Nevertheless, women did not share in the growth of the industry to the extent that men did because they were disproportionately affected by the labor-saving innovations in switching. A recent study (Denny and Fuss, 1983) of employment in Canada Bell between 1952 and 1972 found that the technical change of direct long-distance dialing had a negative effect on all occupational groups studied—operators, plant craftsmen, clerical workers, and other white-collar workers—but that it was greatest for the least-skilled group, operators. The effect of increased output was to increase employment, and the most-skilled workers benefited disproportionately. Ironically, in the United States women were also disproportionately affected to some degree by the implementation of the consent decree agreed to by AT&T and the Equal Employment Opportunity Commission in 1973, which tended to reduce the proportion of females among operators without substantially increasing opportunities for women elsewhere in the system. The number of female operators in AT&T fell from 137,493 in 1973 to 94,586 in 1979; the percentage of females among operators fell from 95.5 to 92.1 in the same period. The percentage of females in all jobs in the company also fell slightly (Northrup and Larson, 1979:46–47).

The 1980 contract negotiated between AT&T and the Communications Workers of America (CWA) established technology-change committees and provided union officials with rights of notice and information with regard to technological changes. The committees are composed of three union and three management representatives from each district and are charged with "the responsibility to develop facts and recommendations" after the company has provided six months' notice of "any major technological changes (including changes in equipment, organization or methods of operations)." (See Chapter 4 for a discussion of the limited role of these committees to date.) The advance-notice feature came into play in 1985 when AT&T announced layoffs for 124,000 workers 60 days in advance. The layoffs resulted at least partly from the breakup of the Bell System (part of a negotiated settlement to a Justice Department antitrust suit) and the resulting increased competition with other telephone equipment and service suppliers. Competing companies enjoyed job growth with increased market share. Women employees have lost or benefited with their company. In the future, employment in some parts of the communications industries may still grow with increases in demand for communications services, but in many parts employment is expected to continue to decline.

In conclusion, the telephone case demonstrates the unexpected and pervasive effects of one innovation and its improvements, the way in which levels of women's employment in one occupation first benefited from and then lost ground with continuing technological change, and the uncertain effects of increased competition and the changing structure of product markets. It also demonstrates the difficulty of assigning causality to observed changes.

WORKERS IN PRINTING AND PUBLISHING

Women have long worked in printing. Artisan production in the U.S. colonial period was a household affair in which wives and daughters were employed, sometimes as typesetters or helpers, sometimes as bookbinders. In 1910 Edith Abbott, a historian of women's work, wrote: "Although the number of women printers has always been small compared with the number of men in the trade, there has probably never been a time for more than a hundred years when women have not found employment in printing offices" (Abbott, 1910:247). In the 1830s with the rise of large-circulation daily newspapers in the burgeoning cities, the relatively undifferentiated industry separated into book and job shops and newspaper publishing. Publishers in the most competitive and technically complex sector—urban newspapers—were the pacesetters of change, much of it organizational. In this period women printers tended to cluster in the book sector, where, if they were typesetters, they did straight text, unbroken lines of type. Newspaper composition, which involved headlines, print in narrow columns that had to be justified, integration of nontype matter and breaks, was done by highly skilled men.

The International Typographical Union, a national trade union, was established in 1850 by the already numerous locals (Lipset et al., 1956:18). In the decade before the Civil War, urban publishers who had introduced the rapidly growing evening newspapers began to hire women who had been trained on small-town newspapers or in book and job shops. During the war, a production bottleneck in composition became apparent. The first typesetting machines, developed then, offered some relief, but they were costly and likely to break down. For newspapers, they were an investment of uncertain return and were not widely adopted (Jackson, 1984:171–174). Instead, publishers accelerated the reorganization of production: for straight text, separated from the complex multitask compositor's job, they hired women who had no previous experience, trained them briefly, and set them to work. Skilled male workers interpreted this policy quite simply as a threat to their economic well-being; they organized and demanded equal pay for women. Employers in turn threatened to reduce male wages to those of women. Nevertheless, male compositors prevailed, partly because most newspaper composing jobs continued to require highly skilled workers; less trained women or boys could substitute for only a small proportion of them (Baron, 1981:32).

The linotype and other automatic typesetting machines were rapidly adopted when better models were introduced in the 1880s. Compositors feared that women would be hired to run them. The manufacturers of the machines, indeed, predicted that cheaper, less skilled workers could be substituted; publishers agreed and tried to do so. They did not succeed, in part because the powerful typographers' national union centralized organizing and collective bargaining. Compositors demanded and won publishers' agreement that machine operators would be apprenticed in the "trade as a whole" (Abbott, 1910:257; Lipset et al., 1956:19; Baker, 1964:44–45). Women continued to learn typesetting outside the apprenticeship system, however, and to be employed setting straight text. Jackson (1984:176–177) concludes: "Mechanizing composition had limited consequences for labor in part because it did not imply an attack upon the skill of the compositor." Sporadic typographical worker association efforts to exclude women seem to have been facilitated also by employer preference for male workers, even though they were more costly. In 1900, 10 percent of all composing room employees were women; by 1940 less than 5 percent were women, and in 1960 only 8 percent (Baker, 1964:172,176).

Overall, the numbers and proportion of women in printing and publishing in general changed little until the 1940s. In the nineteenth century, women worked with boys as paper feeders, although here also, their proportion first declined as faster machines were introduced, and paper feeders functioned as assistants to pressmen; the proportion of females later increased once more, as the job became a dead end rather than a step in a possible male career (Baker, 1964:178).

Women also performed certain specialty tasks, greeting card and paper box printing and binding. Other women in the industry had clerical jobs, or if in the production branch, they were employed in nonunion shops. For many years, printing and publishing was an exceptional case: an industry that, despite employer reorganization and the introduction of technology, remained at core the province of male skilled workers; this outcome must be credited at least in part to the International Typographical Union. However, in the post–World War II period, change in the sex ratio of printing workers accelerated: the percentage of females more than doubled from 13.2 to 28 between 1950 and 1960.

The more recent decades have seen an even more dramatic turn of events in newspaper publishing. In New York and other large cities, the old linotype setting was replaced during the 1960s and 1970s by teletype setting, a process that took mechanically produced punched tape and automatically produced hot metal type; the labor saving was substantial. One operator, tending three automatic linecasters, could produce as much type as seven or eight linotypists (Rogers and Friedman, 1980:3). Innovation continued, moreover, in the form of the "Metroset," a fully electronic machine that does not involve any hot metal type but simply phototypesets electronically produced and laid out images. Reporters and editors now work in front of word-processing screens that show how a page will look; printing has been transformed.

In the 1970s the newspaper typographers' unions negotiated a number of contracts that protected their members' jobs. The relevant clause from the Job Security Agreement (first negotiated in 1974 and renewed most recently in 1984) between the Northwest Typographical Union (No. 99) of Seattle, Washington, and the Tribune Publishing Company of Tacoma, Washington, reads in part:

The Publisher agrees that all of its composing room employees whose names appear on the attached Job Security List will be retained in the employment of the Publisher in accordance with accepted rules governing situation holders for the remainder of their working lives unless forced to vacate same through retirement, resignation, death, permanent disability, or discharge for cause provided.

The agreement provides for exceptions to the job guarantee, such as permanent cessation of the company, strikes, lockouts, acts of God (e.g., flooding, earthquakes). Several similar agreements also provide for paid productivity leaves (leaves in addition to vacation that result from productivity improvements brought about by technological change). This particular agreement also spells out conditions of use for specific innovations and allocates functions between departments and workers.

Newspaper compositors—and printers in general—continue to be predominantly male. Nevertheless, there has been a great increase in women's employment in the industry in the last two decades. Women were 28 percent of the industry's employees in 1960, 33 percent in 1970, and 41 percent in 1980. Total employment in the industry increased by 42 percent from 1970 to 1983; women's employment, by 106 percent (Bureau of Labor Statistics, 1985a:542–543).

Historically, the competition between men and women for jobs in printing went on for the better part of a century, during which time innovation did not displace male compositors. Finally, great labor saving came with new electronic processes, employment growth slowed, and a shift in the sex ratio of workers in favor of women occurred. In this industry, women's employment increased through substitution for men that accompanied adoption of electronic automation.

THE AUTOMATED OFFICE AND ITS WORKERS

The nineteenth-century office was a primarily male workplace. The undifferentiated and potentially upwardly mobile male clerk so vividly invoked by Lockwood (1958) was aided mostly by boys—messengers and "office boys"; office women, copyists or stenographers, were few. Cohn's recent (1985) study suggests that the "clerk" occupation covered a range of positions, many of which did not involve the mobility that earlier analysts imagined. However, the male presence in the office that they described remains valid.

After 1870 the organization and employment patterns of clerical work in the United States changed in distinctive and important ways. Mills (1956:68–69) offers both context and a quantitative estimate of change:

The organizational reason for the expansion of the white-collar occupations is the rise of big business and big government, and the consequent trend of modern social structure, the steady growth of bureaucracy . . . the proportion of clerks of all sorts has increased: from 1 or 2 percent in 1870 to 10 or 11 percent of all gainful workers in 1940.

The acceleration of women's employment in clerical jobs dates from the commercial introduction of the typewriter in 1873 (Baker, 1964:71; M. Davies, 1982). There was a 30 percent increase in the number of mostly female stenographers and typists in the single decade between 1890 and 1900 (Baker, 1964:73). In 1883 the first Burroughs adding machine was put on the market. The number of bookkeepers, cashiers, and accountants increased by 51 percent between 1880 and 1890 and by 60 percent between 1890 and 1900. Although the machines promoted accuracy and speed and thus saved labor, the increasing volume of business meant many more workers were needed. At the turn of the century, women were 29 percent of bookkeepers and accountants. In 1889 Herman Hollerith patented the punched card and counter-sorter, a device that performed calculations, classified cards, and typed out its results. Again, new jobs, many held by women, as keypunchers and machine operators, appeared.

Offices began to be systematically organized, or "socially rationalized," even before the massive introduction of accounting machines and the "Taylorization" that came during and shortly after World War I (Baker, 1964:213, quoting Mills). In the first decades of the century, economic growth promoted both the spread of machines and the reorganization of offices, with increased numbers of women workers. Differentiation and specialization occurred in both of the branches of clerical work: handling words and manipulating data. The occupation of secretary provides a closer look at the clerical occupations that deal primarily with words. The occupations that deal primarily with numeric data are considered below.

SECRETARIES

The secretarial job is found in all industries and in all sizes of offices. In small office settings, where the job involves one person working directly for a boss, substantial changes in the secretary's job often correspond to substantial changes in the boss's job.

As Murphree (1985) points out, the secretarial job is a diverse one; "secretary" tends to be a catchall category for any office worker who performs a variety of tasks that support the work of someone else, usually a manager or professional. Although typing and filing may be the most salient (or observable) characteristics of the job, they actually constitute a small portion of the average secretary's time; while they require a large proportion of aggregate office work time, they are often performed by other office personnel in addition to secretaries. The average secretary performs customized tasks of an administrative and personal nature for the boss; her work flow is unpredictable, as it responds to immediate demands rather than to long-term projects. The duties of a secretary typically vary substantially according to the number of people she works for: the larger the number, the more limited and mundane her tasks are likely to be, as there simply is not time for more customized tasks. Especially in large organizations, secretarial work, like most clerical work, involves much interaction and negotiation with other divisions of the organization (with personnel, contracts, purchasing, sales, travel, and so forth). Manuals of administrative procedures are typically incomplete—if they exist at all—and acquiring information efficiently depends on having well-established informal networks with coworkers throughout the office or firm. In both large and small organizations, secretarial jobs often require a high degree of interpersonal skill.

The authority relations between a boss and a secretary are an important part of the job. Job duties are negotiated individually with a boss and may change with a new boss. The amount of responsibility (and challenge and variety) "given" to a secretary by a boss is highly variable. Loyalty and dedication, as well as initiative and enthusiasm, are important parts of a secretary's job. Secretaries have generally been at the top of the clerical labor force, the best educated and best paid.

The number of secretaries increased enormously with the postwar economic expansion—by 318 percent between 1950 and 1980 (Hunt and Hunt, 1985a). Technological change came to the secretarial occupation slowly, in the form of gradual improvements to the typewriter—electrification, correction function, variable typefaces, electronic memory. These innovations did not involve major reorganization of work (except in the IBM concept of the centralized typing pool linked to mainframe computers, which was not widely adopted) but rather an upgrading of the quality of work that secretaries could produce, perhaps with some time saving. These technologies, and the first introduction of electronic word processing, apparently had little effect on the number of secretaries through 1980. Up to that date, the number of persons whose occupation was secretary increased more rapidly than did the labor force as a whole. In the recession of 1981–1982 the number of secretarial jobs actually decreased for the first time in history. Although growth resumed in 1983–1984, and the number has returned to its former level, two close analysts of the data believe that it is apparent that "secretarial employment growth has slowed dramatically" (Hunt and Hunt, 1985a:3.16). Whether this trend represents reduced demand or a shortage of supply is not clear. Good secretaries are skilled and valued workers, often college-educated. With new opportunities in the professions and management, college women who once might have been secretaries have sought, and found, other jobs.

Murphree (1983) reports that the impact of new word-processing equipment on secretaries differs according to the degree of centralization of work, the spatial arrangements of the equipment, and the type of supervision the spatial arrangements encourage. Production in word-processing centers, where an organization locates all of its word processing at a single site, is often factory-like. The work of manuscript typing is broken up into its components: the pickup and delivery of work, the monitoring and scheduling of work, the entry of words on the keyboard, proofreading, and entering corrections. Work is sometimes rigidly paced and closely monitored. In many instances the word-processing secretaries may have quotas to meet: the entry of a certain number of pages, lines, or even characters. Their work is often monotonous and meaningless, since as word processors exclusively they are not involved in the activities to which the written work pertains (conference planning, billing, sales, research, and so on). The development of centralized word processing also changes the work of those who remain more general-purpose secretaries (those who do not take on one of the new specialized tasks in manuscript processing), because tasks done by the new centralized unit are no longer done by them. If many of their tasks are taken over by new word processors, they sometimes find themselves assigned to a larger number of bosses, and their main function becomes gatekeeping.

In contrast to the centralized pattern, secretaries who work with word-processing equipment in decentralized arrangements generally find the variety, autonomy, and skills of their jobs less affected. They continue to perform tasks they always have: meeting time emergencies, handling unpredictable tasks, and using negotiating and social skills to deal with difficult people. Murphree's research supports the view that word-processing equipment either is making little difference in secretarial work (where it is incorporated into the normal routine, substituting for some typing) or is contributing to the reorganization of work, the subdivision of jobs, and the deskilling of the secretarial job.

Hirschhorn (n.d.) suggests that the traditional secretarial job is becoming increasingly obsolete with the introduction of automation into offices. But he sees it being replaced by different types of paraprofessional jobs, rather than being deskilled. As managers and professionals increasingly do their own keyboarding, those who formerly served as their secretaries take on new duties. New software and the integration of data bases allow the combination of words, graphics, and data in "printing" high-quality reports. According to Hirschhorn (n.d.:1), secretaries may "supervise a document production process, rather than operate within it." They might monitor the organization of work, maintain the equipment and software, seek new applications, negotiate with suppliers and users, and instruct and serve users. Hirschhorn proposes that three paraprofessions may be emerging: the parapublisher, the paralibrarian, and the paramanager. The first supervises the document production process; the second supervises computer-based file and index management—an increasingly important function, since knowledge of how to gain access to the stored information becomes more critical as more is stored; and the third prepares reports or schedules for managers. Hirschhorn's research supports the view that lines between managers and secretaries are blurring.

Although office automation has been occurring more or less continuously for some time, with varying effects on the secretarial job, it is important to realize that offices are by no means fully automated today. Reasons offered by analysts include the diversity of equipment available, the lack of appropriate software, and the cost of both hardware and software. It is also possible that the slow adoption of automation is linked to the fact that some of the new office equipment's capabilities challenge fundamental work hierarchies. The "new office," where similar equipment is used by workers at different levels who share access to integrated information systems, suggests some blurring of the customary lines between managers and support staff; a consequence may be slow or incomplete transformation (Salmans, 1982; Iacono and Kling, 1986; also see section below, "Bank Tellers," regarding an urban bank's introduction of data-processing equipment to its customer service department with the express intention of contributing to the democratization of the office). Moreover, many of the interaction and negotiation functions of secretaries noted above, for example, may not be readily susceptible to automation.

Of the new automated capabilities that are available, word processing is by far the most widely used and seems to be gaining the most rapid acceptance. One *Fortune* 500 company that automated with the goal of doubling the productivity of its salaried staff in the decade of the 1980s estimated that word processing had reached nearly 50 percent of its intended users within the first four years of the program. The company had estimated that the largest demand would be for transaction processing (on-line purchase orders, sales, travel vouchers, and so on), which had already been automated to some degree, and it completed the automation of this type of work early in its decade-long phased changes. In addition to word and transaction processing, other functions undergoing automation in this company were technical computation, data inquiry, business analysis, and electronic mail. Overall, the program has expanded the use of automation: just before its start in early 1979, only 6 percent of the salaried staff used automated equipment; by January 1983, 21 percent did (Alexander, 1983). In this company an effort was made to reach managers, professionals, and technicians as well as clerical workers; however, some of the uses of automation were clearly intended more for one group than for others. One unintended effect of the shift to automation was the subtle realignment of many jobs at all levels as the nature and relative importance of various tasks in any one job changed and as tasks shifted between jobs.

The popularity of word processing no doubt has to do with its relatively low cost, its technical capabilities, and its obvious usefulness as a way to avoid repetitive typing tasks, which constitute a large proportion of office work. Although word processing, like the other automated office functions, has been available on a range of equipment from mainframe computers through minicomputers to stand-alone units, its growth has increased since the advent of the stand-alone. Stand-alone units may be dedicated word processors or more flexible personal computers. The typical large organization office today is likely to have access to and make use of a variety of computerized services on different-sized machines. For example, word processing may be available on a minisystem in a centralized location for major manuscript jobs as well as on stand-alone units or terminals at the secretary's (or professional's) desk for smaller tasks. Data inquiry is likely to be available from a mainframe through a desk terminal or personal computer. The technical composition of offices is in a very fluid state as improved equipment continues to be introduced. The "newest" configuration is a desk workstation that allows access to a large data base, automated filing, data analysis capacities, word-processing and other report formatting tools, computer mail (including automatic calendar functions), telecommunications more generally, and other specific services. By one estimate, 32 million to 38 million workstations will be in place by 1990 with an average investment of $25,000 (Johnson and Rice, 1983).

Office equipment purveyors have made exaggerated claims of productivity increase for much of this new equipment—just as sellers always do. Early estimates of the effects of office automation suggested that secretarial output could increase from 25 to 150 percent with the introduction of word processing. There has been little actual study of productivity before and after installation of word-processing equipment, so it is difficult to verify such claims. It may be that one of the motivations for the adoption of office automation is managers' fascination with new technologies and their exaggerated hopes of potential productivity gains (Iacono and Kling, 1986). After the equipment is introduced, its shortcomings become obvious and its use becomes routinized—with few new, exciting applications being developed. As time goes on and new and better equipment is developed and enters the marketplace, managers become disappointed with their old equipment and possibly replace it with new.

According to Johnson and Rice (1983), who interviewed representatives of 196 organizations by telephone about the implementation of office automation, word-processing systems have usually been brought into organizations with little planning. The initiative usually came from management (90 percent), and often one manager was responsible for the choice of equipment and the implementation process (61 percent). Initial decisions were influenced most by considerations of cost; later decisions to upgrade, replace, or augment were more influenced by factors such as vendor training, maintenance, and ease of application. Training needs were vastly underestimated; little evidence was found of "orderly procurement and installation of equipment, design of jobs, or overall attempts to integrate" (Johnson and Rice, 1983:3). No organization studied in advance the way work and work relations might change. Rather, the innovation was generally thought of as a "horseless carriage," in this case an automated typewriter, that would simply replace one machine without changing anything else. Widespread anecdotal evidence attests to the frequent frustration that occurs when planning and training are inadequate, equipment fails, and tasks are continually altered and shifted without sufficient attention to the consequences.

These common experiences may be compared with those of one distinctive case, in which the managers of an organization deliberately established a planning/implementation team (Yin and Moore, 1984). This team, composed of representatives ranging from top management to secretarial levels, planned and reviewed the implementation of a new office automation system over a one-year period. The collaboration and feedback led to the ability to adapt the new system to the ongoing needs of the organization. Workstations were located and applications and job responsibilities were developed parallel to existing functions, thereby improving the transition from the old to the new system. In addition, the planning/implementation team became a long-term mechanism for considering new applications, the expansion of the original system, and the upgrading of the system in terms of newer hardware.

In summary, during the recent period of widespread adoption of electronic word processing the number of secretaries has not decreased. Although the rate of growth of the occupation has slowed, the relationship between the adoption of new technology and this slowdown is not clear. There has been some specialization and differentiation of functions with the introduction of electronic information processing. Newly specialized workers doing routine word entry are no longer "secretaries" in the broad sense. Several researchers report that the process of introducing microprocessing equipment in large organizations as well as in small offices has been inadequately thought through and planned. Both workers and managers have suffered (or have not benefited to the degree anticipated) as a consequence. Many secretaries have been able to reduce time spent on unpleasant tasks and take on more challenging work; others have found themselves working for more bosses and taking on more gatekeeping as productivity improvements have facilitated doubling up. Finally, it is also possible that the new technology could facilitate secretaries taking on functions, such as financial planning, now reserved to managers. Whether such a flattening of hierarchies is likely depends on the extent to which established organizational relationships are maintained or transformed.

Accountants and Bookkeepers

Mechanical devices have a long history in the data-processing clerical occupations; electronic devices were introduced earlier in these occupations than in word processing. Even before World War I, accountants and bookkeepers became differentiated occupations in the handling of business accounts. The number of women bookkeepers using machines increased, but male clerks tended to resist these changes; some of them lost their jobs to women as a consequence. The greater volume of work and new services that proliferated during the war created still more jobs for women. The Women's Bureau reported that by the end of the war, women were "entrenched as bookkeeping machine operators in customer accounts" (Erickson, 1934, quoted in Strom, 1985:17). Women were being hired instead of men because they made possible (according to one businessman) "a greater volume of business and lower unit cost. Women were paid less than men so cost reduced" (Erickson, 1934, quoted in Strom, 1985:19).

Office employment rose more rapidly in the 1920s and 1930s than the number of machines sold (Mills, 1956:193; Strom, 1985). Apparently, once a certain level of machine adoption was reached, the number of workers again increased. Offices in large firms continued to be reorganized and centralized. Factory-like techniques were applied, as tasks were analyzed, functions separated, and workers arranged in efficient spatial relationship to each other. The depression saw continued feminization of the bookkeeping field with increased use of machines, again the result of cost-cutting by employers. The period also saw the first large-scale organizing, mostly in the CIO affiliates—the United Office and Professional Workers, the United Federal Workers, and the State,

County, and Municipal Workers of America. These unions organized clerical workers in both data-processing and stenography jobs. (Some of these unions collapsed in the postwar period when their leaders refused oaths required by the Taft Hartley Act.)

Another boom in bureaucracy occurred during World War II; again, there was greatly increased demand for calculators, bookkeeping and billing machines, improved machine and work design, and increased female employment. As new opportunities opened or military service beckoned, male workers left bookkeeping.

Mills (1956:204), in a sociological study that is now a historical source, stressed the "factory-like flow of work" in which numeric data were processed and analyzed in the modern office; an increasing proportion of routine jobs (p. 205); and the demotion of "the stratum of older bookkeepers . . . to the level of the clerical mass" (p. 207). He concluded (p. 209): "the new office is rationalized: machines are used, employees become machine attendants; the work . . . is standardized for interchangeable, quickly replaceable clerks; it is specialized to the point of automation."

The 1950s saw the earliest business application of computers that had been developed for scientific calculation and research. Historians of technology disagree about whether the new computers are a continuation of trends in mechanization, a change in kind not degree, or a break. Earlier, one analyst argues, "the pioneering firms had little incentive to put on the market devices which might displace profitable existing data processing machines—mainly punched card tabulators and calculators" (Rhee, 1968:54). By the mid-1950s, however, electronic computers with greatly increased capacity for data storage and rapid calculation were being installed by banks and insurance companies that handled large amounts of numeric data (Department of Scientific and Industrial Research, 1956:42).

The possible adverse employment effect of electronic technology was discussed in the 1960s (Baker, 1964:220): "Observers seem to agree that the displacement of workers will go farthest in the office, and that since office automation aims principally to replace routine, repetitive jobs, women clerical workers may be the largest affected group." Other contemporary observers differed; they saw a greater effect on employment in manufacturing. Baker noted that case studies of several insurance companies that installed computers showed that in some situations there was "considerable upgrading of job content and skill" (p. 225). She concluded, finally, that "an expanding economy will supply clerical jobs for all who seek them" (p. 235).

Indeed the number of bookkeepers did increase, but unevenly, from 1950 to 1980. There were two slow periods, the 1950s and the 1970s, divided by a rapid spurt in the 1960s. After another short growth spurt in the late 1970s, the growth of jobs in bookkeeping has stagnated. Any relationship of this pattern to the introduction of microprocessing equipment would be hard to establish, for growth slowed after 1978, well before wide-scale adoption of microcomputers (Hunt and Hunt, 1985a). The increased use of mainframe computers was undoubtedly important over the whole period.

The data-processing occupations have been highly stratified by sex. Accountants and auditors, the higher-status, better-paying branch, were 90 percent male until World War II and 85 percent in 1950, while women were an increasingly high proportion of bookkeepers and cashiers, 78 percent in 1950. By 1980 the proportion of accountants who were women was 38 percent, while bookkeepers were 90 percent female. In the 1980 census, accountants were reclassified from "professional workers" to "management-related occupations"; this change suggests reorganization of job content and, perhaps, some downgrading of the occupation as the proportion of females in it has increased.

In the first wave of electronic automation, then, women's jobs increased in bookkeeping and accounting both through growth of the industry and through substitution of women for men. It is still too close in time to the very recent wave of automation in bookkeeping to separate its effects from those of general reorganization of tasks in the field and to weigh the relative consequences for levels of women's employment of the technology and the changes in which it is embedded.

Insurance Clerks

Work in the insurance industry had been routinized decades before electronic data processing was introduced. In the back offices of the large companies, functions were broken down into discrete tasks, and paper moved around the office (often rather large ones with many rows of desks) with each person contributing a fragment to the complete activity. Baran (1985) attributes this early rationalization and the division of labor of insurance offices to their similarity to factories—large-scale producers and manipulators of their product, data.

The insurance industry was a leader in adopting mainframe computers. They were used to determine the routing of work and to perform repetitive functions, such as premium transactions and claims disbursements. As computers were introduced, employment grew in two areas: professional-level computer specialists (e.g., programmers and systems analysts) and clerical workers (computer equipment operators and keypunch operators). The logic of the work flow changed very little; the computer simply did some tasks that had formerly been done manually.

During the early introductions of computers, there were varying patterns of effect on numbers of workers. One large company with 4,500 clerical workers in 1960 eliminated 757 traditional clerical positions and added 154 new keypunchers by 1964; managerial and technical staff increased, but since the new employees were primarily male, there was a decrease in the proportion of females of the work force (Helfgott, 1966, cited in Feldberg and Glenn, 1983). A Bureau of Labor Statistics study (1966) of the introduction of electronic data processing in another large insurance company found "relatively few lay-offs" as a direct result. This result was believed to have been possible because of the high turnover of the tabulating and calculating machine operators, most of them young women. However, there were fewer entry-level jobs for young high school graduates because the skills required for first jobs in the new areas increased.

Advances in both hardware and software design led to the computerization of some aspects of underwriting and rating in which the computer, using algorithms, would make decisions formerly made by underwriters and raters. But until the mid-1970s, these changes occurred within the established organization of production. In the late 1970s, however, several factors produced pressure for more far-reaching changes. Inflation, high but unstable interest rates, deregulation accompanied by increasing competition from other parts of the financial industry, and systematic applications of computer and information technologies created both the demand for and the possibility of reorganizing the insurance industry. This reorganizing was done in two ways: by installation of electronic data processing in horizontally integrated systems and by feminization.

The new automation modified the work process. While the early use of computers was based on and reinforced the fragmentation of jobs within established hierarchies, the newer applications of information technology integrate fragmented tasks to create new jobs, often while eliminating old ones. An example of this reorganization comes from property and casualty insurance, where back-office jobs with the carrier are being eliminated in favor of data entry and complete processing by local sales agents (Appelbaum, 1984:10):

One major insurer . . . is developing computer programs that assist in underwriting and rating of these commercial lines, and (sales) agents are being trained to use these programs . . . highest volume agents (about 10 percent of the sales force) [are supplied] with intelligent terminals linked to the central computer . . . and to printers capable of issuing a policy. When the procedures are fully implemented, all clerical processing will be eliminated for sales generated by these agents. Policy typist, rater, and underwriter's assistant—jobs which provided career paths from clerical to lower level professional jobs as underwriters—are disappearing . . . and lower level jobs as insurance professionals are rapidly declining as well.

"Personal lines" insurance—standardized products that can be mass-marketed through direct mail and television advertisement—provides a second example. In one company the unskilled jobs of mail handling and filing were automated by 1980; policies were entered directly into the computer, and customer inquiry clerks could key into the system to answer questions about coverage immediately. The work process was extremely fragmented, however. Depending on whether they came by mail or telephone, customer inquiries were answered by different people with different job titles. To save money in an intensely competitive market, the company undertook steps to reintegrate the work process. By 1983 a new, highly skilled clerical position had been designed. Customer service representatives handle sales, have access to the computer program that assesses risks and to the rating program, explain rating procedures to customers, answer customer questions, and respond to complaints by telephone or mail (Applebaum, 1984:12). Here, job integration ranges across functions, combining sales with processing and both with customer service/client relations. A third model—the use of computers in insurance in the early pattern—coexists with the integrated models; computers assist professionals and clerks in traditional jobs who produce insurance policies that cannot be standardized.

These examples offer interesting contrasts. In the first one, some clerical duties and professional activities were folded over into sales work to create a new kind of sales position. The remaining professional activities, e.g., underwriting exceptional cases and overseeing the work of the agents, were at the highest level. Some clerical jobs—e.g., typing reports, processing exceptional cases, and entering data—remain. In the second example, clerical duties have been expanded to include some sales and professional activities, supported by computerized systems. Here the sales force and professional positions were greatly reduced in number.

The model of work organization varies by the kind of business a firm does, but in all of the varieties of work organization "job categories have become more abruptly segmented while avenues of mobility between them have been sharply reduced" (Appelbaum, 1984:12). As a result of the automation of underwriting and claims estimating for standardized insurance policies, career ladders from skilled clerical to insurance professional positions have been eliminated or shortened. The gap between these types of jobs has widened at the same time that the gap between entry-level clerical jobs and skilled ones has become formidable. In the aggregate, skill requirements for clerical workers have increased because computers are doing the most routine tasks. At the same time, however, jobs at all levels have become increasingly dead end.

Feminization occurred as the insurance business reorganized. Between 1970 and 1978, women were hired in 231,000 of the 259,000 new jobs created in the insurance industry. Over the next four years women were hired for 113,900 jobs, while men were hired for only 20,300. Women increased their proportion in clerical and nonclerical (professional, managerial, and technical) jobs that were formerly preponderantly male. Minority women also increased their proportion in insurance jobs as automation proceeded, but most of them are in back-office jobs in data-processing centers, typing pools, and filing departments (Baran, 1985). Overall employment in insurance from 1958 to 1984 is estimated to have increased 1.72 times, slightly above the average (1.67 times) for the economy as a whole (Hunt and Hunt, 1985a). At the same time, productivity growth meant that sales grew faster than employment. In the aggregate, the number of women in insurance has increased both absolutely and relative to the number of men during automation.

Based on her case study of 55 insurance companies, Baran (1985) concludes that although the discrimination barrier was lowered to permit women to move up from clerical to professional positions in insurance, a new structural barrier has blocked access. Although women can enter professional insurance positions through certification, noncertified women will not be able to move up to management and professional posts through on-the-job experience. The new skilled clerical positions that they can enter, especially if they have college degrees, became increasingly women's jobs in the 1970s as women doubled their percentage in such jobs as insurance adjusters, examiners, and investigators. These are the very jobs Baran believes will be directly affected by the automation of claims procedures. A different projection was offered in an aggregate statistical study (Hunt and Hunt, 1985a) that argues that adjusters and investigators, like secretaries, although supported by computerized systems, interact directly with other persons as part of their jobs; hence, they are unlikely to be replaced by machines. As evidence, they point to the continued rapid increase of insurance adjuster jobs in 1981–1982, when much of the clerical sector stagnated or declined. Black women, who entered clerical work only in recent decades, remain primarily in the more traditional, less skilled jobs that are increasingly at risk of elimination through automation. However, although small numbers are involved, black women are overrepresented, compared with other administrative support occupations, as insurance adjusters and investigators.

Overall these changes in the products and the production process of the insurance industry have brought about changes in employment and occupational patterns. In contrast to the steady increase of employment from 1960 to 1980, from 1980 to 1982 there was a marked slowdown in growth, and, for some kinds of insurance, an absolute decline in the insurance labor force. It remains to be seen whether this slowdown is a temporary reflection of an economywide recession or a more permanent outcome of the changes in work organization. What does seem permanent is a shift in occupational structure. The number of lower- and middle-level clerical positions has declined by being automated out of existence or combined with other jobs. Examples include keypunch operators, bookkeepers, and file clerks, occupations that declined by approximately 16,000 workers between 1970 and 1978 (Appelbaum, 1984:Table 2). Skilled clerical positions are expanding, but they seem to be increasingly dead-end jobs, with little opportunity for advancement. For example, insurance adjusters and examiners and computer operators grew by 69,000 workers between 1970 and 1978 (Appelbaum, 1984:Table 2). Some lower-level professional positions have also disappeared—again through automation and the redesign of jobs— while those remaining often require high-level professional credentials and managerial ability. The question of mobility between occupations in the insurance industry has been acutely posed by the reorganization of the work process and redefinition of jobs that have accompanied the adoption of new equipment (Appelbaum, 1984). Women still predominate in the insurance work force and are an increasing share of the professional employees; no employment crisis has emerged. Locational shifts of offices, from urban to suburban areas or to rural or different regions, a trend reported by some observers, may yet produce problems for workers unable to move, but it is too early to trace or evaluate its effects (Baran, 1985). Furthermore, the increased emphasis on college education may further narrow opportunities for less educated black and white urban working-class women.

As in several of the other industries examined here, one worker response has been to use collective bargaining as a means to make the relations between automation and the existence of jobs, and also their conditions, a contractual matter. Although it covers only some employees in a small Syracuse, New York, claims office, the November 1984 contract between the Service Employees International Union (SEIU) and the Equitable Life Assurance Society of the United States is far-reaching. The contract calls for 60 days' advance notice of the introduction of "automation or equipment that will result in (a) a reduction of employees, (b) substantial changes in any employee's jobs, or (c) a change in job classification." This contract also deals with such issues as employee monitoring by computer; provision of retraining to affected employees; and a job security provision requiring that the Syracuse office be kept open, with the bargaining unit maintained, with at least the number of employees as at the time of the signing of the agreement. (After unionization, but prior to the successful negotiation of this first contract, the company had threatened to close the office.) In addition, the SEIU contract goes into great detail with regard to the use of video display terminals (VDTs). The contract calls for rest breaks, glare reduction, provision of detachable keyboards and adjustable chairs, regular eye examinations, and provision for assignment to non-VDT duties for pregnant workers who request such temporary transfer.

Bank Tellers

Banking, like insurance, pioneered in the use of computers for processing and storing information (Ernst, 1982). Banks—then employing mostly men— once offered service primarily to businesses. As service to individual clients increased, banking mechanized even before computers, first with mechanical calculators and later with electromechanical check-sorting equipment. Proportionately more women were employed in new, less skilled jobs (Strober and Arnold, 1985). Nearly fully automated check processing began with the use of machine-readable encoding of the bank and account from which the check originates, along with the encoding by the receiving bank of its code, the account, and the amount of the check. Magnetic ink character recognition (known as MICR) began in the 1950s; standards for acceptable printing ink and code location and so on were adopted in 1958. By 1963, 85 percent of all checks were being encoded. This development, of course, sped the adoption of computers for record-keeping for other aspects of checking accounts.

Tellers were provided with computer-based equipment in the early 1960s, so data entry began at the point of the transaction. Today, with an automatic teller machine (ATM) (facilitated technologically by the microprocessor) for selected simple transactions, a consumer enters the information directly into the computer, and the machine provides the cash requested, accepts the deposits, or performs funds tranfers.

The bank credit card, another innovative service to individual consumers, was introduced on a national scale in the late 1960s; the widely used and accepted credit cards expanded consumer debt and centralized its servicing, replacing many store charge-account systems. Through the 1970s the number of checks cleared in the United States grew at a rate of about 7 percent per year; the volume of credit card transactions increased even more rapidly (Gurwitz and Rappaport, 1984–1985).

In addition to services to individual consumers, banks are also involved in the management of their own assets and in facilitating cash management by their business customers. Higher interest rates make electronic funds transfer especially important for cash management. Many businesses and other large organizations can communicate directly with banks to keep track of their assets and transfer their funds as desired. Banks now often clear their net transactions with each other electronically. The electronic transfer of funds through home computer-bank linkage is available to depositors in some large banks, but this procedure is not yet common.

The banking industry has undergone changes under the impetus of forces similar to those that have affected the insurance industry: inflation, high but unstable interest rates, deregulation, and competition from other financial intermediaries. The most striking consequence is the wide range of new products and services offered by banks and the seemingly inexhaustible demand for financial services of all kinds. As automation has proceeded, it has served to improve banks' competitive position both as it reduced unit costs and made expansion of product and service lines possible. Without these tremendous productivity gains in banking, the checking system would almost certainly have faltered; the quality of services would have declined while their prices rose, leading to slower growth (Ernst, 1982). As it is, there is no question that technology contributed to the growth in demand for the product and led to increased employment, even though many of the innovations were also labor saving.

Over the 30-year period from 1950 to 1980, bank tellers were the sixth fastest growing clerical occupation. Female employment in this occupation increased 12.9 times between 1950 and 1980. Current Population Survey (CPS) data from 1972 to 1982 show bank tellers as the second fastest growing clerical occupation, with a 103 percent increase in female workers in that period. These same data, however, show an end to growth, with the number of employed bank tellers smaller in 1982 than in 1981. Between 1983 and 1984, the CPS data show no increase. (The changes made in occupational categories between 1982 and 1983 make it impossible to compare the data through 1982 with those from 1983 and subsequent years.) Overall employment in the banking industry from 1958 to 1984 grew 2.7 times, a rate that is far above that of all 105 industries in the economy (1.7 times), and even above that of the top 10 clerical-employing industries (2.2 times) (Hunt and Hunt, 1985a).

These changes have brought about shifts in the types of workers needed, as well as in their numbers. The need for back-office clerical workers has been reduced, while demand for highly trained sales personnel has increased. Technically sophisticated sales workers will continue to be increasingly important because of increased competition among financial institutions, continuing introduction of new products and services, and the growth in larger, more sophisticated institutional customers. The teller's job is somewhere in the middle of these trends. As consumers perform the simpler transactions through ATMs, human tellers are likely to deal primarily with more sophisticated requests.

Sales might provide an opportunity for upward mobility for tellers (Hirschhorn, n.d.). Many tellers now deal primarily with exceptional situations (certified checks rather than routine deposits and withdrawals) and the sale of bank services (cash management accounts). The implications of this shift for tellers' mobility opportunities in banks are not yet clear, however, because there may be countervailing tendencies. Osterman (1984) suggests that the internal labor markets of banks, too, are changing, making career mobility within the bank more difficult. The sales job is becoming a highly skilled one, more dependent on formal training and more tolerant, even encouraging, of interbank mobility. At the same time, on-the-job training and the institution-based internal career ladder for bank tellers have tended to disappear. Nevertheless, women's participation in upper-level bank jobs has increased more rapidly than their share in all management jobs. Although some of this shift may reflect job title inflation, some of it reflects increased management opportunities created in the recent rapid expansion of the retail side, affirmative action, and the increased willingness of employers, customers, and employees to consider women in these jobs.

Even if the jobs do continue to provide internal career mobility, employment in the first place will be more difficult because entry-level jobs are likely to require more education. This may exclude less well educated women (and men), who are likely to be disproportionately minority, poor, and urban (Noyelle, 1985).

A study of the introduction of an integrated data base system in the customer service department of a large New York bank (Center for Career Research and Human Resources Management, 1985) demonstrates one way that leading industry institutions can offer a model for the introduction of new technology. The bank's stated policy in advance of any change was "no layoffs." The manner of change was planned with positive outcomes for workers—reduction of stress and upgrading—among its goals. There was redesigning of both the physical setting of the department that enhanced the working environment and of jobs that promoted professionalization or skill upgrading of the clerical workers involved. Some clerical workers experienced upward mobility. The job redesign also "democratized" the office; this change was appreciated by clerks, less so by middle managers who felt that democratization reduced or leveled their status. Time will tell whether these contrasting perceptions of the outcome are reconciled. The clerical work force was reduced by 14 percent in the changeover; presumably those workers who lost their jobs moved elsewhere in the bank or left voluntarily.

Growth of employment in banking is likely to slow. The advances in automation have taken place primarily in the retail part of the business, the most labor-intensive sector and the sector in which banks had been the most actively soliciting business. Regulation held interest rates down and encouraged banks to compete in other ways for retail business, by opening branches and charging low service fees (which kept the cost per transaction to the consumer artificially low). With gradual deregulation over the past five years (it is expected to continue), bank branch opening declined and service charges were raised; banks now have less incentive to cater to small-scale retail trade (Gurwitz and Rappaport, 1984–1985). With further automation (more ATMs, point-of-sale terminals, and home banking) likely to eliminate much of the paper that has until now been generated and processed (even though processed electronically for the most part), labor needs in retail banking will probably decline. The assets of banks also are now growing more slowly and erratically, especially relative to their high and steady growth prior to 1978. Thus, in banking, both future growth patterns and employment levels are uncertain, but employment growth on the retail side is very likely to slow.

In summary, increases in women's employment as bank tellers accompanied preautomation business and organizational changes. During the recent rapid growth of the banking industry, women's employment increased as automation contributed to growth of demand for new products and services. Women's move into bank management occurred in the same period. Recently, however, employment growth has slowed. Continued reorganization has changed conditions for entry-level positions and internal mobility, with the potential of closing off opportunities for less educated women. Future changes in the pattern of demand, in the organization of work, and in automation are likely, and those changes will continue to affect employment.

RETAIL CLERKS

The retail industry in the United States has undergone striking changes in the twentieth century. A continuing process of technological change, much of it outside retailing itself, coupled with general social, cultural, and economic change, has transformed retailing from an industry dominated by single-proprietor neighborhood stores serving highly localized markets to one dominated by national chains and holding companies, most of whose member stores serve broad regional markets. Some of the changes outside of retailing that have been important in this transformation include the development of nationally recognized brands and standardized packaging by manufacturers (which began in the late nineteenth century); advances in freight transportation, which improved distribution of national brands to local outlets; the development of suburbs and automobiles, which made regional shopping centers possible; advances in general office procedures, including record-keeping and billing; the development of new forms of consumer credit, which increased the ease and amount of consumer credit; and new media (radio, movies, television), which promoted mass consumption as well as advertising messages for specific products. Retailing, which is directly dependent on consumer needs and preferences, has generally responded rapidly to these changes and initiated innovations of its own.

Among the most prominent changes since World War II has been the development of chain, discount department stores, such as K-Mart. From a handful of such stores in the 1950s, the number grew to 1,300 in 1960 and to almost 7,400 by 1977. During the same period, chain department stores, such as Sears, Roebuck and Company and J. C. Penney, also expanded rapidly, and holding companies, such as Federated Stores, expanded their holdings to include prominent specialty stores as well as locally distinctive department stores. Chain and discount stores now handle 89 percent of the department store trade; department stores in general have about 80 percent of the general merchandise market. Concentration in this industry continued to increase even after the period of rapid expansion of the chains: between 1967 and 1977 the top 32 firms increased their market share from 75 to 87.5 percent of all sales, and within this group the top five increased their share from 49.8 percent to more than 60 percent (Bluestone et al., 1981:48). The most rapidly growing sector of the industry, however, is discount chain stores. And since the energy crisis of the 1970s, mail-order catalog sales have also increased dramatically.

In other sectors of retail trade, such as hotels, restaurants, and grocery stores, the trends are similar to those in general merchandise retailing: increasing scale through chain operations, more reliance on recognition of national "brand" names (Safeway, Howard Johnson's, McDonald's, Hilton, Sheraton), greater use of advertising, and increased use of automated equipment. For example, the 20 largest supermarket chains (those with more than 100 stores) increased their share of the grocery store market from 27 percent in 1948 to 41 percent in 1977. The introduction of scanning technology in supermarkets (uniform product codes read by scanners at the cash register) was increasing at a rate of about 10 percent of all stores per year, meaning that its introduction would be virtually complete by the end of the 1980s. Cash register scanning eliminates marking prices or price changes on products. Data recorded in the process are used to control inventory, automatically reorder products, schedule workers' hours, and allocate shelf space. The new machines also reduce checkout time (Burns, 1982).

The retail industry as a whole is highly competitive; consumer income is limited, and most areas of retail trade are not expected to grow more rapidly than disposable income, although some sectors can experience high growth rates by capturing changing preferences. Even when there is market saturation, a "new concept" in retailing can capture business from competitors. For example, a declining part of the average family budget is spent on groceries, but an increasing part is spent on eating out, and within the grocery sector, market share can be captured by the super supermarket with its vastly expanded offerings including prepared, ready-to-eat food, and the warehouse store with its curtailed product selection and lower prices. Interestingly, both the super store and the warehouse store in the grocery trade employ fewer workers per dollar unit of sales than the standard supermarket, and the standard supermarket is losing market share to the newer forms (Burns, 1982).

A recent study of changes in the department store part of the retail market suggests that computer-based technological change has been essential to the growth of multiple-store chains (Bluestone et al., 1981). Computer-based inventory control, ordering, and warehousing, as well as the product scanning codes and intelligent cash registers, created economies of scale that made expansion possible. Without automation, the amount of record-keeping required to operate at such scale would probably have been prohibitive. Mass media advertising, another technique that benefits from economies of scale, has also facilitated the growth of chains and added to their relative advantages over independent stores. Indeed, Bluestone and his colleagues (1981) argue that advertising has allowed retailers to reduce the numbers of higher skilled sales workers and substitute less skilled and lower-paid sales help. The salesperson no longer needs to sell the product; that has been achieved by advertising.

The distribution of occupations in retailing between 1950 and 1980 documents the shift toward less skilled cashier jobs. In 1950, 4.3 percent of all jobs were in retail trade (excluding cashiers), and cashiers constituted another 0.4 percent of all jobs. Between 1950 and 1970, the labor force in retail trade grew from 2.5 million to 2.9 million, representing an annual growth rate of less than 1 percent; between 1970 and 1980, the growth rate for the entire decade was just over 1 percent (Bureau of the Census, 1964:Table 201, and 1984b:Table 221). Overall, employment in retail trade grew more slowly than total employment.[1] The number of cashiers, on the other hand, more than doubled between 1950 and 1960, and has continued to grow rapidly since: between 1960 and 1970 their number increased by 73.4 percent, and between 1970 and 1980 by 87.0 percent. By 1980 cashiers constituted 1.8 percent of the total labor force and 34 percent of all retail trade employees (Bureau of the Census, 1984b:Table 221).

The shift toward the use of cashiers strongly favored female employment. Cashier was a predominantly female job already in 1950, and over the subsequent three decades it increased from 76.5 to 83.0 percent female. Thus, most of the new cashier jobs (approximately 1.2 million) went to women.[2] While other retail jobs grew more slowly than that of cashier, new employment opportunities again favored women; the proportion of those jobs held by women grew from 48.9 percent in 1950 to 58.7 percent in 1980 (Bureau of the Census, 1964:Table 201, and 1984b:Table 221). However, sales of higher-priced items, generally accompanied by higher pay, continued to be male-dominated in 1980. Men claimed a disproportionate share of sales jobs in motor vehicles and boats, furniture, radio, TV, hi-fi equipment, and appliances, while women heavily dominated the counter clerk and apparel sales jobs (Bureau of the Census, 1984b:Table 221).

While the shifting composition and content of retail jobs have been accompanied by expanding job opportunities for women, the quality of those jobs has deteriorated in several respects. Salespersons are playing a less central role than previously, even in the higher-priced stores that provide relatively more customer service, such as Bloomingdale's. Commission sales are rapidly disappearing, and standard hourly wages only slightly above the minimum wage are becoming the norm. Along with the use of less skilled sales workers, part-time employment has increased: when it is no longer necessary to have skilled sales workers who know the products, scheduling part-time workers to meet daily and weekly fluctuations in demand—and to reduce costs—becomes possible.

---

[1] Because occupational classifications changed between census years, comparisons across the entire period are difficult to make. However, comparisons between adjacent census years (1950–1960; 1960–1970; 1970–1980) are generally possible. There was a major change in classification in the 1980 census. The 1970 and 1980 data compared here use like categories but cannot accurately be compared with previous years.

[2] To compare cashier jobs across census years, data were taken from Hunt and Hunt (1985a), who adjusted census data for purposes of consistency.

Extending store hours to improve competitive position also contributes to stores' hiring of part-time workers. Scheduling is facilitated by computerized data from intelligent cash registers that can be used to identify fluctuations in demand. In the New England region, which led in the development of discount stores, it is estimated that fully 75 percent of all department store employees work part time. One national discount chain estimated that only 15 percent of its work force is full time (Bluestone et al., 1981:83). Much of the employment is seasonal as well as part time; only one-third of department store employees work year round; more than one-third work only one-quarter of the year, usually around Christmas.

The increased use of less skilled and part-time workers has also contributed to making the job of department store salesperson less likely as a career; career opportunities for the floor salesperson have virtually disappeared. These contrasting trends reflect the reorganization of retailing just discussed: sales personnel no longer "sell," but sales still occur; cashiers handle the transaction.

The department store industry is now characterized by enormous turnover of young workers rather than by more stable employment of older workers. The proportion of department store workers in New England who were under 25 years old increased from 30 percent in 1958 to more than 50 percent by 1970. The percentage of the department store labor force in New England that had at least three years' tenure fell from more than 42 percent in the 1950s to less than 32 percent in the 1970s, while the percentage who had worked less than one year increased from less than 20 percent to more than 30 percent (Bluestone et al., 1981:84). In most years between 1957 and 1975, more than 40 percent of the labor force left the industry each year. With such high turnover rates, much hiring occurs even in years of declining employment. Obviously, training is minimal; otherwise this staffing strategy could not make economic sense. Pressure to reduce labor costs is large, since competition is stiff and labor costs are a relatively large proportion of total costs. Payroll costs range from 11 to 14 percent of total sales in the most standardized parts of the industry (chain and discount stores) to 16 to 27 percent in the less standardized parts (holding companies, independents, and specialty shops). Because of the competition, unions have generally been unable to prevent pressure on wage rates, and little of the work force is unionized. The young age of the workers, the high rate of turnover, and the large proportion of part-time workers all appear likely to make more widespread unionization difficult.

Unionization has had a somewhat stronger foothold in food retailing, largely because skilled meat cutters in supermarkets were organized. In fact, workers in some of the unionized department stores (Bradlee's) became so as a result of mergers with supermarket chains (Stop and Shop). The intense competitive pressures in food retailing led to a merger between the retail clerks and the amalgamated meat cutters unions in 1979, creating the United Food and Commercial Workers, with 1.2 million members. Nevertheless, the strength of the union is declining, especially in supermarkets, which had been its strongest section. Technological change in meat packing has resulted in the near elimination of skilled meat cutters in supermarkets; some of the meat now arrives in the stores boxed, labeled, and ready to be sold, while most of it is in boxes of cuts (rather than whole carcasses) that need little further cutting before being packaged and sold; since 1974 nearly 4,000 meat cutter jobs have been lost in four cities alone—Boston, New York, Chicago, and Los Angeles (Burns, 1982). Without the high end of the wage scale to "anchor" other wages, wages of supermarket workers are falling in relative terms, and unions have had to accept two-tier wage bargaining, with new entrants on lower wage scales (with smaller future increases) than present workers.

Technological changes have directly contributed to the changes in retailing, particularly its increased scale and use of part-time workers, both of which have been facilitated by computerized record-keeping. Other important technological changes that have transformed the industry over a long period include general advances in communications and transportation and increased use of advertising. The discount branch of the industry, with its lower labor needs, is growing most rapidly. "Decreased reliance upon labor is a by-product of concentrated ownership" (Bluestone et al., 1981:50), which continues to increase. Employment of cashiers has increased, providing a large pool of new, low-skill, low-wage jobs for women—jobs with little stability and few career opportunities. Other retailing employment declined or grew slowly. The basic competitive structure of the industry no doubt contributes to this outcome, for it leads to a strong need to keep labor costs to a minimum.

NURSING AND NURSES

Like the other industries discussed here (telephone communication, printing and publishing, insurance and banking, and retail trade), the health industry is undergoing fundamental structural change. The restructuring of the health sector is a result of changes in insurance programs, increases in the relative supply of medical doctors, growth in the number of profit-making hospitals and clinics, changes in the age structure of the population, and altered preferences for types of health care. Together with technological change, these factors influence the work performed and the number and qualifications of the people required to do it. Employment growth in the health industry over the past several decades has been well above average. It is projected that employment growth in the health industry will continue to outpace that of the U.S. economy as a whole into the 1990s, but at a slower rate than that experienced during the 1970s.

Nurses comprise the largest group of health professionals in the United States. The numbers of both registered nurses and licensed practical nurses increased substantially between 1970 and 1980 (Bureau of the Census, 1984b). Throughout this century, the nursing occupations have both grown steadily and undergone continuous change. Until the depression, most of the trained nurses in the United States—in contrast to nurses in Britain and elsewhere—worked in private home nursing and community health rather than in hospitals. This fact enabled them to argue that hospital training was inadequate for their practice and to establish and consolidate a variety of educational programs, including university training for registered nurses (C. Davies, 1980). For decades nurses have been split into different groups, according to training and accreditation, place of employment, and specialization within nursing.

The most recent national sample surveys and estimates indicate that of the close to 1.7 million registered nurses, 1.4 million actually practiced their profession in 1983. Adjusting for part-time workers, this amounts to 1,174,200 full-time equivalents—600 employed or 502 full-time equivalent nurses per 100,000 population. Just under one-third of employed registered nurses had a baccalaureate or higher degree. Registered nurses constitute the ninth largest occupation for women and the highest-paid predominantly female occupation. In the 1980s two-thirds of the registered nurses employed in nursing worked in hospitals. The second-largest group practiced in nursing homes (8 percent). Both of these groups exhibited the highest estimated growth rates in the employment of registered nurses since 1977, 39 and 27 percent, respectively (American Nurses Association, 1985). The growth rate in hospital employment is expected to slow and to be accompanied by changes in types of positions and by requirements for more education or experience (Sekscenski, 1984; American Nurses Association, 1985).

Although their education and training is shorter and less professionally oriented than that of registered nurses, licensed practical or vocational nurses (LPNs or LVNs) also require accreditation. In 1983, of an estimated 781,506 LPNs and LVNs, 539,463 (69 percent) were employed in health care. More than half of this group, 57.6 percent, worked in hospitals, followed by 22.5 percent in nursing homes, and 9.1 percent in physicians' or dentists' offices. Trends toward demands for higher formal education of nurses, particularly in hospitals, are also indicated by the fact that the total number of LPNs and LVNs has not grown as rapidly as that of registered nurses (Sekscenski, 1984:12, 17). And after a continuous increase in full-time equivalents of LPNs in hospitals for many years, their numbers decreased from 1982 to 1983 (American Nurses Association, 1985:46). Unemployment rates for LPNs have usually been higher than for registered nurses (in 1983, 5.4 and 1.6 percent, respectively), but lower than for the population as a whole (American Nurses Association, 1985).

There have been only small increases in the percentage of males among nurses. The proportion of males was roughly the same among both registered nurses and LPNs and LVNs, 3 percent in 1983 (American Nurses Association, 1985). Male nurses are generally better paid than female nurses and occupy a disproportionate share (one-fourth) of the supervisory and administrative positions in nursing (Jacobson, 1983:50).

More LPNs and LVNs than registered nurses are ethnic or racial minorities, reflecting less access to higher-level education by minorities, but whites dominate both occupations, constituting more than 90 percent of registered nurses in 1980 and close to 80 percent of LPNs and LVNs in 1983 (American Nurses Association, 1985; National League for Nursing, 1985). In 1980 an estimated 88.7 percent of minority registered nurses were employed in nursing, compared with 75.7 percent of nonminority nurses.

Nursing has been strongly affected by technological developments in medicine and the biosciences. Both in hospitals and in the community, nurses have continually had to adjust to innovations in medical technology. Computers and information technology, more generally, are increasingly linked to other forms of technology in use in the health services.

Computers were first introduced in the U.S. health care system in the 1950s. They were by-products of computerization in other sectors of the economy, not specifically designed for health services. The first applications occurred in large hospitals, primarily for administrative routines such as inventory control, billing, and payroll. Few health professionals were involved in the development and refinement of computer systems in hospitals. Laboratories in large hospitals were the first clinical area in which computers were applied successfully. Medical statistics were also an important application, parallel to administrative computing. Only occasional attempts were made to integrate clinical and administrative systems.

Knowledge and learning are central aspects in the diffusion of technology (Rosenberg, 1976). In the first phase of computerization of hospitals, not much transfer of knowledge about the new technology, let alone about its interconnections with organizational change, occurred. Expertise with computers and computer use in the health services was located either outside hospitals, with vendors and shared-service computer companies, or in separate computer departments, which only large hospitals could afford. The hospitals that started early to use computers for accounting and patient records steadily developed more complex procedures. At the same time, technology and concomitant specialization increased the necessity for coordination and communication. One of the main functions of nursing became the coordination of a whole array of specialist medical competences in the service of patients (Fagerhaugh et al., 1980). While a few individuals in the 1950s envisaged the possibility of automating selected nursing activities and records, computing procedures were cumbersome, and reprogramming was costly. Staffing requirements for managing and improving computer systems and the complexity of the structure of information and communication in hospitals were grossly underestimated (Ball and Hannah, 1984).

Gradually, mainframe computers were replaced by minicomputers and microcomputers, keypunching machines by direct data entry. Improvements of computer systems notwithstanding, many forms of "resistance to" or "interference with" computerization on the part of doctors, middle management, nurses, and clerical and technical personnel are reported from the 1950s into the 1980s (Startsman and Robinson, 1972; Henskes and Kronick, 1974; Watson, 1974; Dowling, 1980; Counte et al., 1983). Frequently, the organization of the implementation process is cited as a cause for resistance or reluctance: lack of timely information, loss of power and control, insufficient attention paid to the privacy of patient information, and so forth. Overselling by vendors and unrealistic expectations of health care personnel led to lack of involvement in improving systems by personnel theoretically able to do so. Changes in traditional procedures of practice are often mentioned as a reason for resistance (Ball and Hannah, 1984). Doctors, for example, persisted in giving direct orders or handwritten notes to clerks, while nurses often continued to write most of their notes by hand (Lievrouw, 1984). In spite of—or perhaps partly because of—computerization, administrative work in hospitals and in the health sector in general increased considerably, and with it the number of clerical workers. By 1983 the health industry employed 1.2 million workers in clerical occupations, more than double the number employed in 1970 (Sekscenski, 1984).

Beginning in the middle of the 1970s, there was a marked increase in the interest of doctors and nurses in computerization and other information technologies. This change resulted both from accumulated experience with the new technologies and from a growing realization of the potential and the feasible adoption of increasingly useful techniques of microtechnology and data base management. Medical informatics became a specialty in medicine, with its own journals and conferences. An extensive variety of specific applications were developed during this period, ranging from hospital and laboratory information systems using specialized diagnostic and patient monitoring, to signal analysis and image processing, to educational and research applications. Many of these technologies integrate knowledge from different medical and nursing specialties, but their use leads again to the formation of new groups of specialists. The changes influence the employability of nurses and make frequent retraining necessary. Depending on how the technologies are organized, they can also constitute sources of stress in the work situation, both in relationship to patients and in the nursing team (Jacobson and McGrath, 1983; Ball and Hannah, 1984).

The monitoring of births and of patients in coronary, neonatal, or postoperative intensive care are examples. Initially, computers were used to analyze electrocardiograms. Data were presented on paper trace records and later on video display units (VDUs) in alphanumeric and graphic form. Watching machines was a tedious, strenuous nursing task, so alarms were combined with the computers and were programmed to alert personnel to deviations from normal. With the further development of the technology of representation of physiological processes and the reduction of noisy signals, technical experts saw total automation as a possibility. It has been suggested that fully automated systems could lead to a reduction in nursing staff time as well as in the medical/technical knowledge necessary on the part of nurses. Automated monitoring, combined with other technology, could result in continuous recording over extended periods, increased automatic storage and analysis of records, and automatic generation of suggestions for treatment and automatic administration of medication (Fox, 1977).

The diverse patterns in the introduction of patient monitoring devices in different countries suggest that—as in other areas—one cannot speak of "the effect" of the new information technology on nurses. Effects depend on choices regarding the extent of automation, training of personnel, their reliance on the data and their discretion in interpreting them in conjunction with direct observation of patients, the distance of the equipment from the patient and the rest of the ward, and so on. A strong commitment on the part of both nurses and physicians to the importance of intensive personal nursing at the bedside and to support for the patient's family seems to underlie decisions about many monitoring systems that purposely have not been totally automated. Increased authority of nurses combining different forms of observation of patients and increased consultation among members of health care teams are also reported as possible outcomes of the introduction of patient monitoring (Medical Informatics Europe, 1982, 1984; Ball and Hannah, 1984; Child et al., 1984).

Today nurses are more often consulted when information technology is introduced into hospitals and other health care settings than they previously were. Nurses are also among the developers of special nursing applications of computer and information technology, including video and telecommunication systems. Manufacturers and vendors have employed nurses to improve existing systems. IBM, for example, has underlined the importance of user involvement at Duke University in adapting its patient care system (PCS) to local hospital needs not necessarily envisaged by the designers: participants in the implementation process at individual hospitals are nurses and other health care personnel as well as computer specialists (Light, 1983). In addition to improvements in hospital administration systems, laboratory systems, and patient monitoring systems as seen from the viewpoint of nurses, one now finds special computer applications for nursing practice and administration in hospitals and community health settings, for basic and continuing nursing education, and for nursing research. Telemedicine, a system that allows practitioners to function without on-site physician coverage, through bidirectional cable television, is being tried in community clinics staffed by nurses (Cunningham et al., 1978). In contrast to earlier computer-based systems with implicit models of the work of nurses, several of the most recent specific nursing applications are based on explicit philosophies of nursing. Examples of such systems range from problem-oriented patient care systems (McNeill, 1979) to systems for clinical decision support to be used in the education and practice of nurses (Grobe, 1984; Ryan, 1985).

The possible employment effects of such systems are, however, hardly spelled out. As is the case for other expert systems, there are possibilities for raising the professional standing and knowledge of one's own group as well as possibilities of diffusion of knowledge to groups with less formal education. The latter effect can lead to replacement or curtailment of growth of the "expert" group through employment of groups with less formal training. Systems planned as labor-saving devices might turn out to demand increased input of labor of another kind, either more or less professional, in this case either more or less directly involved in nursing practice. Whether groups of workers perceive the introduction of new technology as a threat to be averted or as a challenge to further their own interests will depend partly on their knowledge base both about their own field and about the technology in question. Professional education and accreditation alone are not sufficient for such a knowledge base. Both formal and informal networks are necessary for gathering information, comparing experiences, and contacting designers and vendors.

In the case of physicians and nurses, professional organizations and international contacts have been starting points for the activities of individuals and small groups specializing in computerization in their fields. Nurses interested in computers and information technology have been able to use their general professional journals for this purpose; they have also established communication and publication channels of their own. An increasing number of books and articles are available to nurses who want an introduction to computing. There are "how to" guides complete with addresses of network contacts, calculations of cost-effectiveness for a nursing administrator, and analyses of data base systems for a nursing computer specialist. There have been many national and international conferences on medical informatics in the 1980s with sections on computers in nursing. In 1982 the nurse members of the American Association for Medical Systems and Informatics (AAMSI) formed a professional specialty group for nursing within the association. Nursing educators in the United States may affiliate with the health education special interest group within the international Association for the Development of Computer Based Instructional Systems. Informal computer interest groups within the national nursing associations have existed for some time. Recently, the American Nurses' Association constituted a Council on Computer Applications in Nursing. In the spring of 1985 the National League of Nursing formed its National Forum on Computers in Health Care and Nursing, a "membership group designed to advance computer technology in the nursing community." Much cross-fertilization of ideas for improving information technology in nursing occurs in international forums. Since its first meeting in Stockholm in 1974, the International Medical Informatics Association (IMIA) has held meetings triannually, with special working groups for nursing education. In 1982 an international working group on nursing was formed as part of the IMIA.

The historic experience of nursing and nurses demonstrates once more the interrelationships between technical change and social relationships. Hospital personnel, including nurses, have become actively involved in the introduction of electronic innovations in patient care. Nurses have organized, through collective bargaining and other means, to participate in the development and implementation of new technologies. Employment effects for nurses so far have been small, because of nurses' knowledge and types of jobs, coupled with generally strong growth in health sector employment. The high level of education of nurses has no doubt facilitated their participation in implementation of new technology and contributed to their job security in the face of change. Comparison with European experiences shows, once again, that the same technologies can be used with very different effects. In the United States, very recent change in federal government rules affecting Medicare reimbursement (the Diagnosis-Related-Group [DRG] basis for fee payment) has produced shorter average hospital stays; some job loss for nurses and other hospital workers has occurred in 1984 and 1985. Other sectors of health care are still growing, and some shifts in employment to those sectors (e.g., health maintenance organizations, nursing homes, offices of nonphysician providers, home health care agencies) are likely to occur. Again, it is too early to evaluate these effects and, in particular, to observe their relationship to technology as separate from more general social and political decisions.

CONCLUSIONS

Like the telephone, the new telecommunications and microprocessing technologies facilitate many product and service activities. As more information is available in electronic form, it can be transmitted around the country and the world almost instantaneously. The space and time compression of modern telecommunications may have the same mixed—and sometimes conflicting—effects on social relations and on geographic dispersion as did the telephone. Similarly, microprocessing has been related both to the growth of new services (in banks and insurance) and the decline of others (sales help in retailing). More generally, the flexibility of the new white-collar technologies suggests that their uses and effects will be influenced by concurrent social changes and by social decisions rather than by any inexorable technological determinism.

What common strands can we extract from these long-term and more recent changes associated with technological innovations in the various cases described above? What problems emerge for further consideration? First, as the longer-term historical examples demonstrate, changes in equipment and in the organization of work occur in combination with other structural change in product and labor markets. In each of the cases discussed, technological change has occurred along with fundamental changes in industrial structure. Identifying the specific impact of technological change is therefore difficult.

Second, technological change plays varying roles in economic growth. Necessity is sometimes the mother of invention (when an innovation meets a perceived need—often called induced technological change), but invention is also sometimes the mother of necessity (when it creates demand by bringing forth new possibilities that people had not known they would want—called autonomous technological change). The telephone and the automobile are quintessential examples of the latter. And in insurance and banking, the ease of calculation and communication caused by the new information-processing technologies led to enormous growth in the products and services that could be offered. Unforeseen effects occur as much internally in an innovating firm's labor market as in its product market. A new technique is often first seen as a simple replacement for an old process, but when it does not quite fit with the old way of doing things, new ways of doing things may evolve and unforeseen ways of using the new technique may develop.

Third, consequently there is great diversity in the way organizations use technical advances—with varying employment effects. In some kinds of production (printing and meat cutting), recent technological and organizational change has been very labor saving, and worker displacement occurred when demand increases were not large. In the banking and insurance industries, rapid growth has prevented displacement, but if the rate of growth slows dramatically, there may be similar effects. In other kinds of work (secretarial and nursing), however, the variety of the tasks and the social relations on the job have led to little labor displacement, and little is likely in the future; the personal relationships that are part of these jobs are not amenable to automation, although many of the tasks are.

While the uses of technology are social choices, those choices are often constrained by such factors as competition and labor force availability. Some businesses may be more constrained than others because of such factors as their market position, the degree of competition they face, the extent of product differentiation they can create, their maturity, the amount of labor they employ relative to other inputs, and the stage of major technological change they may be undergoing. Given this diversity, it follows that workers' and managers' interests will sometimes diverge. It is difficult to find, in the past or the present, instances of workers' introduction of technological change; these decisions are generally management's. Sometimes cooperation in training and learning among workers themselves can speed adoption and facilitate the efficient use of new techniques. Nevertheless, workers must generally adapt as best they can. It should be remembered, however, that managers are also workers—middle management in particular may be threatened by new technologies that have implications for changing the nature (and number) of their jobs.

Fourth, with respect to the skill levels required for given jobs, there are countervailing tendencies of increased complexity and of greater simplification and standardization. Some technical developments in telecommunications and financial services—for example, satellite and information technologies—require more sophisticated personnel. In other areas, standardization akin to industrialization in the last century is occurring, particularly in service delivery, with likely tendencies toward lowering average skill levels. Examples here include data entry and mass mailing of standard life insurance policies.

Fifth, the recent rapid increase of women in the labor force has been a correlate of recent change in the service sector, just as an earlier transfer of women from household production to wage work outside the home accompanied changes in manufacturing (textiles, garment making). Most of the recent increase in women's employment in services has not been a result of competition with men and replacement of male by female workers: the feminization of the office, for example, occurred long before the introduction of microprocessing and telecommunications technologies. Rather, substantial increases in the demand for labor have been met by women. Employment growth, then, has been related to overall expansion of service industries, both absolutely and in comparison with the manufacturing sector. Both the recent and past changes demonstrate that the decline or disappearance of specific jobs is offset by growth elsewhere. In the past, employment declines in some areas produced varied effects: turnover and natural attrition were sufficient to avoid involuntary unemployment; some involuntary separation occurred; and growth in other occupations offered new jobs for some, probably most, workers. Recent increases in structural unemployment, however, raise concerns about the future. So, too, do the unequal effects of such unemployment on recently hired people, less educated minority workers, and handicapped workers.

Sixth, with respect to the issue of employment quality, in the past some employers were forced by circumstances to improve the quality of work—for example, by too great turnover (as in the early years of telephone operating). Worker mobilization and demands sometimes achieved improvement or job protection. Other employers were motivated to improve work quality by a genuine wish to improve conditions for their employees. Such policies have been most often implemented in periods of expansion and prosperity. Today, there is heightened consciousness of quality of employment issues among both employers and workers in a less expansionary economic climate. The possible consequences are not clear.

Women workers today have more opportunities generally, higher and more continuous labor force participation, and greater expectations. Their turnover rate now closely resembles that of men workers. In some occupations women are prospering; in other occupations, their jobs are at risk; in still others, the quality of their employment is low. The next chapter provides an overview of present levels of employment and occupation structure and considers future prospects for women workers.

3

Effects of Technological Change: Employment Levels and Occupational Shifts

What are the implications of innovations in telecommunications and information processing, and associated changes in the organization of work, for the availability of jobs to women, particularly over the next 10 years? The previous chapter suggests that these innovations have varying effects on the quantity of jobs in different occupations and industries. Much may depend on how they are implemented and on underlying economic and social conditions. If the new information technologies do adversely affect clerical employment, women will be disproportionately affected. Over the next 10 years, will there be significant displacement of present clerical workers? Will there be sufficient clerical jobs for all those who seek them? Any attempt to look at future effects is, of course, limited; at best, prediction is uncertain, an informed guess. This chapter seeks to reduce that uncertainty. The context of this inquiry is one of substantial controversy in the existing research literature and substantial concern expressed in the media.

Some studies predict large relative job losses for clerical workers. For example, in a major study based on an input-output matrix for the United States with moderate economic growth and with two scenarios of diffusion of computer-based innovations, Leontief and Duchin (1984) predict that employment growth in clerical occupations will be very small relative to growth in the total labor force. With the most rapid diffusion, they projected total employment to grow from 89.2 million person-years in 1978 to 124.1 million in 1990 and to 156.6 million in 2000; clerical employment, however, was projected to grow from 15.9 million in 1978 to only 16.7 million in 1990 and 17.9 million in 2000. Hence, under their assumptions, clerical workers would decline from 17.8 percent of the labor force in 1978 to 13.5 percent in 1990 and to 11.4 percent in 2000. Furthermore, Leontief and Duchin project that if workers continued to train for the mix of skills required in 1978, it is possible that 5 million clerical workers would be unemployed by 1990. Of course, job growth would occur elsewhere: in professional, service, craft, and operative occupations. Although a number of criticisms have been directed at this study (see Hunt and Hunt, 1985a,b; Kraft, 1985), it systematically illustrates one end of the range of predictions.

Several economists have noted that the job-creating aspects of new technologies are more difficult to predict than the job-displacing aspects are to observe. In testimony before the House Committee on Science and Technology, Nathan Rosenberg (1983) pointed out that although technological change virtually always involves occupational shifts, it need not involve overall job loss; he suggested that current U.S. employment problems are more the result of macroeconomic policy than of technological change. Similarly, Eli Ginzberg (1982), introducing the special technology issue of *Scientific American,* emphasized the job-creating nature of technological change. Without technological advance, economic growth slows and employment declines can occur, especially in industries that fail to innovate.

A case in point with respect to clerical work is provided by the finance industry, a heavy user of automated information processing and a classic example of technological change that did not result in an employment decline. Between 1972 and 1982, clerical workers as a proportion of all workers employed in the finance industry decreased from 46.1 to 43.9 percent (Hunt and Hunt, 1985a), indicating that it became possible to produce any given output with relatively fewer clerical workers than previously. Office automation may have contributed to this relative reduction in clerical workers. During this same period, however, total employment in the industry expanded substantially—from 3.9 million to 5.3 million, an increase of 35.9 percent, about 1.5 times the rate of employment increase in all nonagricultural industries (Bureau of Labor Statistics, 1985b)—in part because the new technologies facilitated the provision of new and better services to consumers at lower cost. Thus, although clerical employees comprised a smaller *proportion* of the work force in the finance industry in 1982 than in 1972, the *total number* of clerical workers employed in the industry rose considerably. Indeed, the increase in clerical jobs in the finance industry over the decade was 37 percent, considerably higher than the 28.8 percent increase in clerical employment for the economy as a whole (Hunt and Hunt, 1985a).

PROBLEMS IN EMPLOYMENT PROJECTIONS

Underlying Factors

Of course, no one can be certain whether the pessimistic view, the optimistic view, or something in between will prove most accurate for clerical employment over the next 10 years. Four major underlying factors affect this future, some of which can be better predicted than others: (1) the general performance of the economy overall and the associated general employment picture; (2) specific policies that affect women's employment opportunities; (3) changes in the supply of labor; and (4) developments in the available technology. These factors are considered briefly here, in reverse order.

The technology that is likely to influence employment in the next decade either has already been developed or its likely parameters are fairly well specified. As described in Chapter 1, the ease of processing and transmitting information is increasing rapidly as the cost falls. Some uncertainty exists about the rate at which innovations in telematics will diffuse throughout the economy. In the panel's judgment, diffusion will not accelerate over the next 10 years: deliberate rather than headlong speed seems likely. The keyboard will remain the main mode of data entry; the new integrated workstation will remain unavailable to all but a few; networking between systems will remain a problem. Of course, unforeseen applications of known technology can alter a situation dramatically. These technologies do have the potential to contribute very greatly to productivity increases, but they remain relatively untapped. The productivity gains that are possible can be applied in several ways: to develop new products and services, to improve quality, or to cut costs. Only the third way is likely to have substantially negative effects on employment levels, and those will occur only if the lower costs (and prices) do not induce increased demand.

The rate of growth of the labor force—i.e., changes in the supply of labor—is obviously an important factor in its adjustment to technological change. If, for example, labor force growth slows considerably while the rate of economic growth does not change, the economy can more readily absorb workers who may be displaced by technology. If, however, labor force growth remains high, workers displaced by technology will be more likely to remain unemployed along with new entrants and reentrants to the labor force. The people who will enter the labor force during the next decade have already been born, and recent trends in labor force behavior can provide a guide to future behavior. Again, however, unforeseen change is possible. Throughout most of the postwar period, for example, forecasters consistently underestimated women's labor force participation. Public and private policies in the areas of child care, transportation, training, flex-time, and so on can change basic structural factors that influence people's labor force participation. It is important to note, too, that the education and training of current workers and new entrants to the labor market affect the ease with which they can respond to changing labor market conditions. (Future labor supply is examined in greater depth below.)

Specific policies that affect women's employment opportunities directly, such as enforcement of equal employment opportunity laws, can have a major impact on the availability of jobs for women. Even if the number of jobs available in occupations and industries that have been dominated by women fall or experience slower growth in the next decade, if women's opportunities increase elsewhere in the labor market, the effect of the shrinkage of traditional employment for women would be mitigated. Currently there is some uncertainty about the equal employment opportunity enforcement effort. Women made substantial progress in integrating many occupations, particularly in the last decade (see Blau and Ferber, 1986; Reskin and Hartmann, 1986), and it seems likely that some progress will continue.

The general performance of the U.S. economy is perhaps the most important factor in assessing future job opportunities for women (or men). If economic growth is rapid and demand for labor is high, all those whose jobs become outmoded by technological change will almost certainly be able to find new ones. But if overall economic performance is sluggish, replacement jobs will be hard to find, and the costs of adjusting to job loss that is due to technological change will be high in both economic and human terms.

Future economic performance is difficult to predict because it depends so heavily not only on national economic factors but also on the outcome of political processes and international economic developments, such as the volume and nature of trade, the strength of the dollar, and the success of OPEC in controlling the price of oil. Among important public policies that affect the general health of the economy and the overall demand for labor are fiscal and monetary policies that influence interest rates, the deficit, and inflation and tax policies, tariffs, and other statutory incentives and disincentives that influence the development and use of innovations and so alter capital/labor ratios. All these policies affect the development of new technologies and the rate at which they are used. And government policies, in turn, are influenced by alterations in political and social commitment to providing full employment.

The panel made no attempt to predict future economic growth, but the discussion in the rest of this chapter implicitly assumes modest growth rates over the next 10 years. In the panel's judgment, this assumption is a reasonable basis on which to gauge the effect of technological change on women's employment. But it is possible that the economic future could be vastly different, and technology could play a different role in that future. If economic performance deteriorates and economic pressures on employers intensify, the productivity gains made possible by the new technologies are likely to be applied to cost cutting, and substantial technological displacement might occur, along with the cyclical unemployment that would result from poor overall economic performance.

Whatever the impact of technological change and economic growth on the levels of employment, their impact on occupational shifts is less ambiguous. Technological change and economic growth invariably generate shifts in demand for workers in various occupations. Clearly, a healthy economy will contribute to the ease of adjustment to these shifts.

Data Problems

Another source of uncertainty in discussing the employment effects of technological change is the inadequacy of the available data for studying the connection between technology and employment. The data are inadequate for assessing the impact of technological change on both the quantity and quality of employment.

To assess changes in employment opportunities or outcomes with any degree of confidence in the generalizability of the results requires data from a representative sample of jobs in a local or national labor market, with jobs grouped into occupational categories on the basis of some standard occupational classification scheme. But none of the existing occupational classification schemes—e.g., those of the Bureau of the Census, the Bureau of Labor Statistics (BLS), the *Dictionary of Occupational Titles,* or the Standard Occupational Classification (SOC)—was designed with the assessment of technological change as an important consideration. These schemes are also inconsistent with one another. Moreover, they are updated periodically (the decennial census classification, for example, is revised for each new census) without systematic attention to the way technological changes have redefined the task content of jobs.

A specific example illustrates the difficulties. Suppose one wanted to determine how widespread the use of word processing was by 1980. In principle, this is the sort of question it should be possible to answer from 1980 census data. It is impossible to do so, however, because the 1980 census classification has no separate category for "word processors." Formerly they were included in the category "typists." For the 1980 census, because word processors work with computer technologies, the Census Bureau decided to reclassify them with keypunch operators, as does the SOC. This change was also implemented in the monthly Current Population Survey (CPS), but not until 1983. The change in the SOC is a dramatic one, affecting major occupational groups as well as more detailed categories such as typist or keypunch operator; 1980 data are now not generally comparable with earlier censuses. The Bureau of Labor Statistics' periodic reports on the labor force would be of no greater help, because they are generally based on the Current Population Survey (which has too small a sample size for reliable data on the numbers of workers who use word processors) and because the classification also does not identify "word processors" separately. Furthermore, none of the classifications in use would identify workers who use word processing as an auxiliary part of their jobs—for example, writers, editors, managers. Thus, there is no direct means of tracing the adoption and use of this new technology. Of course, industry statistics on the sale of stand-alone word processors and on the sale of word-processing programs for general-purpose microcomputers would allow some basis for an estimate, but such statistics can reveal nothing about which categories of workers use word processors, how many workers share each workstation, and so on, or how the organization of work has changed in response to the introduction of such equipment. The fundamental problem is that information on technology is not linked to information on workers in currently available data (see Hunt and Hunt, 1985b).

The inconsistencies across time, data sets, and agencies cause difficulty even in analyzing categories such as secretaries or clerical workers, let alone word processors. Although BLS uses "census categories" in the household data collected in the Current Population Survey, it uses other classification systems in other data collection efforts. The area wage surveys, for example, rely on the categories used by the employers being surveyed. (Some of these may even identify word processors separately.) A third data collection effort, the Occupational Employment Statistics (OES) program, the biennial surveys of establishments that form the basis for the occupational projections, uses yet another classification scheme. Consequently, the numbers that are developed from the data sets sometimes differ widely for the same occupational category.

For example, according to OES data, there were 2,797,000 secretaries in 1984 (Silvestri and Lukasiewicz, 1985); according to the annual averages from the CPS monthly data published in *Employment and Earnings* (January 1985), there were 3,935,000. Although the OES classification system is different from the CPS, the difference probably does not stem from differences in the definition of "secretary" between the two systems but from differences between self-reporting (CPS) and employer reporting (OES). For another example, OES data show a total of 18,716,000 administrative support workers in 1984 (17.5 percent of the civilian employed labor force of 106.8 million), while the CPS total is 16,722,000 (15.9 percent of the total of 105.0 million). In the 1980 census, the count was 16,851,000 (or 18.5 percent of the labor force).

To address some of these data problems, the panel commissioned a study on recent and future trends in clerical occupations from H. Allan Hunt and Timothy L. Hunt (1985a) of the Upjohn Institute. They developed a reasonably consistent set of occupational categories for clerical workers across the 1950–1980 censuses. To look at more recent trends, they use CPS data for the 1972–1982 period, and they sometimes report 1983–1984 CPS data separately because they are not compatible with earlier data. In looking at projections, they use the BLS projections based on OES data. Their estimates form the basis for the panel's discussion later in this chapter.

Some of the same concerns apply to the assessment of employment quality, which is discussed in Chapter 4. There are several systems of inquiring about the quality of work life: the *Dictionary of Occupational Titles,* the Quality of Employment Survey, and numerous disparate surveys on job satisfaction. None has emerged as the clearly dominant method for measuring quality, and none is likely to do so. Moreover, there has been no systematic attempt to link employment quality to specific technologies used on the job, although occasional private attempts have been made, e.g., the survey of women who do and do not use video display terminals that was taken of the readership of women's magazines (9-to-5, 1984a). There is at present no way of knowing how many jobs are affected by technological changes, much less what the effects are.

A statistical system that does not distinguish between typing on a typewriter and data entry on a word processor or keypunch machine is inadequate as a basis for determining the extent of technological change in the office. The point is not limited to office jobs, of course, but pertains equally to many categories of jobs affected by technological changes. The lack of an adequate data base requires that one maintain a healthy skepticism regarding research on both the quantity and quality of the employment effects of technological change.

This chapter next turns to further examination of the supply of labor, particularly women's labor. It then reviews recent trends in clerical employment as a base from which to understand future possibilities. The following section reviews several aggregate projections of demand for clerical labor, illustrating a "most plausible worst case scenario" that the panel believes places a lower limit on the likely demand for clerical employment in 1995. Employment projections for specific subfields of clerical work and likely occupational shifts and changes in skill requirements are also discussed. The conclusion attempts to link information about supply and demand to assess the impact of technological change and other factors on employment levels in clerical jobs, currently the dominant source of jobs for women.

THE SUPPLY OF WOMEN WORKERS

Labor Force Participation Rates

The future supply of women workers is a function of the number of women and the rates at which different groups of women participate in the labor force. The number of women is determined by the fertility that prevailed 16 or more years earlier, by women's current mortality rates, and by the net balance of women's immigration and emigration. Of these factors, of course, the first is generally dominant. Women (and men) who will be at work in 1995 have already been born. Barring unforeseen developments in mortality or immigration, we can project fairly confidently the number of women of working age in 1995 (see, for example, Bureau of the Census, n.d.). In addition, knowledge of fertility, mortality, and migration permits one to project current and future age distributions, which is important because women's labor force participation rates vary by age.

Current projections by the Census Bureau indicate that the working-age population will continue to grow, although at a slower rate than in the recent past.

The slowdown is primarily a result of the fact that the baby boom generation has come of age and already entered the labor force. The population cohorts reaching working age over the next 10 years will not be as large.[1] If labor force participation rates remained constant by age, the projected average age of the labor force would also rise.

The rates at which women will participate in the labor force are, however, much more difficult to project. Two factors contribute to the difficulty: the variations in participation rates among groups of women and the rapid increases in participation rates among almost all women. The rates at which women participate in the labor force have historically varied according to women's ages, their potential wage rates, marital status, husband's income if married, level of education, and presence and age of children. The rates have also differed by racial and ethnic group. Labor force participation rates have historically been highest for women aged 20–24; for unmarried or divorced women; for well-educated women; and for women without children, especially young children, at home (Bureau of the Census, 1985:Tables 671, 673, 675).

The rates for black women have historically been higher than those for white women, and white women's rates have been somewhat higher than those for most women of Spanish origin (who may be of any race). In 1984 the labor force participation rate for black women was 55.2; for white women it was 53.3; and for women of Spanish origin it was 49.6 (Bureau of the Census, 1985:Tables 660 and 670). But there can be marked differences within a group: for example, in 1984 the Cuban women's participation rate was 55.1, higher than that for any other group of Spanish origin and higher than the rate for all white women. (For a description of racial and ethnic differences in women's rates, see Sullivan, 1978.) And although differences in rates among groups of women have persisted, there has been a pronounced, and largely unanticipated, increase in the labor force participation rates among all groups of women. While no single reason for this increase can be identified, many recent economic, social, and cultural changes are conducive to it.

Among demographic changes, lower levels of fertility and the greater use of child care have made it easier for mothers of young children to hold jobs; at

---

[1] Between 1985 and 1990, the population over the age of 16 will increase by about 4.8 percent for both men and women. The rate of growth will slow between 1990 and 1995 to around 3.8 percent, with the male population growing at a slightly higher rate of 3.9 percent (calculated from Bureau of the Census, n.d.:43, 53, 63, 73; middle series of projections). During the five years from 1995 to 2000, as the children of the baby boom enter the labor force, a growth of 4.4 percent in the female working-age population and 4.6 percent in the male working-age population is projected. In 1985 the working-age population is estimated at 95.3 million women and 87.7 million men; by 2000 the projected population will be 108.3 million women and 99.9 million men. Principally as a result of recent fertility trends, the median age of the population will rise from 32.7 for women and 30.2 for males in 1985 to 34.9 for men and 37.7 for women in 2000.

least one study suggests that further improvements in child care would encourage even more such mothers to work (Presser and Baldwin, 1980). Later ages at marriage and at first childbearing have enabled many women to gain greater job seniority and so perhaps increased their commitment to work outside the home.

Among factors related to financial need, inflation and husbands' unemployment or the erosion of husbands' real wages may have encouraged some wives to enter the labor force or to remain in it. Financial pressures are also a factor in decisions by divorcees or widows to reenter the labor force. Moreover, increases in marital instability in the past several decades may have encouraged some women to maintain their work roles as a source of financial independence.

Institutional changes, including legal action against sex discrimination and sexual harassment in education and employment, greater equity in pensions and other fringe benefits, and greater availability of maternity leave, may also have played a role in the higher participation rates. Cultural change, coincident with the rise of the feminist movement and reflected in the changed aspirations and expectations of women, may be especially significant. As more women work, employment becomes the norm; there are changes in attitudes toward women's work and the role of work in adults' lives. These changes have probably encouraged more women to remain in the labor force.

Finally, some analysts have argued that shifts in labor demand have increased opportunities for "women's jobs" (Oppenheimer, 1970). Although it is difficult to demonstrate that women entered or stayed in the labor force because of the availability of jobs, it is significant that very large numbers of the same kinds of workers who had traditionally held certain jobs arrived in the labor market just at the time of highest demand for those jobs.

Projections of Labor Force Participation Rates

Given the number of influences that might be affecting women's labor force participation rates, it is very difficult to make assumptions about trends in the rates. Hence, projections of future labor supply do not have a firm basis. While the number of women of working age is well known and relatively easy to project, past projections underestimated the rapid increases in the rates and therefore in women's labor supply. Because no one can predict whether the participation rates will continue to increase or at what rate they will change, projections of women's labor supply are uncertain. By contrast, projections of men's labor supply have been relatively accurate because a basic assumption has been reliable: after completing their education, very high proportions—more than 75 percent—of all working-age males will remain in the labor force until retirement age. This assumption holds almost without regard to men's marital status, education, or any other characteristic except health status, al-

TABLE 3-1 Actual and Projected Labor Force Participation Rates for Women Aged 16 and Older by Race, 1970–1995

Year	All Women	Whites	Blacks[a]
1970	43.3	42.6	49.5
1975	46.3	45.9	48.9
1980	51.5	51.2	53.2
1984	53.6	53.3	55.2
1990[b]	56.6	n.a.	n.a.
1995[b]	58.9	58.4	62.7

[a]In 1970, refers to "blacks and other races."
[b]Projected rates.

SOURCES: Data for 1970, 1975, 1980, and 1984 from Bureau of the Census (1985:Table 660); data for 1990 and 1995 from Fullerton (1985:Tables 2 and 3).

though black men have lower labor force participation rates (by about 5 to 10 percent) than white men at all ages.

The best available current projections are those of BLS. Table 3-1 presents women's labor force participation rates between 1970 and 1984 and projected rates for 1990 and 1995. These projections are probably the most responsible professional estimates of labor force participation rates, but it should be noted that rates rose so quickly after 1970 that labor force projections for 1985 were surpassed during the 1970s. The estimated size of the female labor force in 1990 would be 55.5 million and in 1995 it would be 59.9 million; in comparison, in 1984 the size was approximately 49.7 million.

Table 3-2 provides the same information for men for the same time period.

TABLE 3-2 Actual and Projected Labor Force Participation Rates of Men Aged 16 and Older by Race, 1970–1995

Year	All Men	Whites	Blacks[a]
1970	79.7	80.0	76.5
1975	77.9	78.7	71.0
1980	77.4	78.2	70.6
1984	76.4	77.1	70.8
1990[b]	75.8	n.a.	n.a.
1995[b]	75.3	75.8	69.5

[a]In 1970, refers to "blacks and other races."
[b]Projected rates.

SOURCES: Data for 1970, 1975, 1980, and 1984 from Bureau of the Census (1985:Table 660); data for 1990 and 1995 from Fullerton (1985:Tables 3 and 4).

TABLE 3-3 Size of Civilian Labor Force, Proportion Female, and Actual and Projected Growth Rates, 1970–1995

Year	Civilian Labor Force (millions)	Proportion Female (percent)	Labor Force Growth[a] (average annual rate of change)		
			Civilian Labor Force	Male Workers	Female Workers
1970	82.8	38.2	—	—	—
1975	93.8	40.0	2.5	1.9	3.5
1980	106.9	42.5	2.7	1.8	4.0
1984	113.5	43.8	1.5	1.0	2.2
1990[b]	122.7	45.3	1.3	0.8	1.9
1995[b]	129.2	46.4	1.0	0.6	1.5

[a]Annual rate of growth from the preceding date.
[b]Projected data.

SOURCE: Fullerton (1985:Table 1).

Although the overall male labor force participation rate is projected to be higher than the overall female rate in every year, the male rate is projected to decline slightly, largely because of earlier retirement. Steeper declines occurred among both black and white males between 1970 and 1984 than are expected to occur from 1985 to 1995. Tables 3-1 and 3-2, taken together, strongly suggest that men will account for a declining proportion of the labor force. Summary projections, shown in Table 3-3, show an increase in the proportion of the labor force that is female and consistently higher growth rates in the female labor force. However, growth of both the male and female labor force is expected to decline after 1985. For the 1970–1984 period, the number of women in the civilian labor force grew at an annual rate nearly twice as large as that anticipated between 1985 and 1995.

PROJECTIONS OF AGE-SPECIFIC RATES

Along with a general increase in women's rates of participation at most ages, some changes in the patterns of age-specific labor force participation are being projected for the next 15 years. For most of the postwar period, women's labor force participation rates peaked at ages 20–24, dropped at ages 25–34, and rose again at ages 35–54. The drop in rates at ages 25–34 was attributed to withdrawal from the labor force for childbearing. In 1970, 57.7 percent of women aged 20–24 were in the labor force; the rate dropped almost 13 points, to 45.0 percent, for women aged 25–34 and then rose 6 points, to 51.1 percent, for women aged 35–44.

In every year since 1970, the percentage of women in the labor force at ages

20-24 has risen, and the drop in rates between ages 25 and 34 has declined. In 1975 the rate for the latter group dropped 9 points, from 64.1 to 54.9 percent; by 1980 the decline was only 3 points, from 68.9 to 65.5 percent. By 1984, the decline was less than 1 point, from 70.4 to 69.8 percent. Projections for 1990 no longer include a decline at ages 25-34 (Bureau of the Census, 1985:Table 660).

In addition to this change, the peak age for women's labor force participation is expected to shift. By 1990, labor force participation for women is expected to peak at 78.6 percent for ages 35-44, rising in 1995 to 82.8 percent for that age group. Although these cross-section rates do not indicate changes in behavior of the same women over their lifetimes, the patterns they suggest are essentially borne out in data about successive cohorts. Table 3-4 shows both cross-section rates and cohort effects (shown in the steps going down the table). Each cohort was less likely than the one before to experience dramatic declines in labor force participation rates during the childbearing years, so that the most recent cohort experienced no decline at all. The members of each cohort, as they age, increase their participation in the labor force, and each successive cohort has higher participation than the one before.

The projected pattern would resemble the male pattern of labor force participation much more closely than have earlier female patterns. The 35-44 age group for men has typically had the highest labor force participation, with rates well over 95 percent. Among males, a striking change has been projected in age pattern—increased withdrawal of older workers from the labor force. In 1970, 83 percent of men were in the labor force at ages 55-64; this proportion is projected to drop to 64.5 percent by 1995. Participation among males over the age of 65 is projected to be about one-half of what it was in 1970, with a decline from 26.8 to 13.3 percent. By contrast, very little change is projected in the participation rates for older women workers: in 1984, 41.7 percent of women aged 55-64 were in the labor force, along with 7.5 percent of those over 65.

OTHER FEATURES OF WOMEN'S LABOR FORCE PARTICIPATION

The increased similarity of the labor force participation of women and men over the life cycle corresponds to an increase in female attachment to the labor force. Women are working more and more consistently over the life cycle. One measure of attachment is the turnover rate, the extent to which workers in the labor market at any one time leave during the year and are replaced by others. If the same workers stayed in the labor market all year long, the turnover rate would be zero. For women workers, the turnover rate has fallen rather steadily from 32.1 percent in 1957 to 12.7 percent in 1983; for men, the turnover rate fluctuated modestly around 8.0 percent between 1957 and 1977 and after 1977 fell to 4.5 percent (Blau and Ferber, 1986:Table 4.3). Although there is still

TABLE 3-4 Labor Force Participation Rates of Women, 20 Years and Over, by Age, Annual Averages, Selected Years, 1955–1985

Year	Birth Cohort	Age Group									
		20–24	25–29	30–34	35–39	40–44	45–49	50–54	55–59	60–64	65–69	70+
1955	1931–1935	46.0	35.3	34.7	39.2	44.1	45.9	41.5	35.6	29.0	17.8	6.4
1960	1936–1940	46.2	35.7	36.3	40.8	46.8	50.7	48.8	42.2	31.4	17.6	6.8
1965	1941–1945	50.0	38.9	38.2	43.6	48.5	51.7	50.1	47.1	34.0	17.4	6.1
1970	1946–1950	57.8	45.2	44.7	49.2	52.9	55.0	53.8	49.0	36.1	17.3	5.7
1975	1951–1955	64.1	57.0	51.7	54.9	56.8	55.9	53.3	47.9	33.3	14.5	4.8
1980	1956–1960	69.2	66.8	64.1	64.9	66.1	62.1	57.8	48.6	33.3	14.7	4.6
1985	1961–1965	71.8	71.4	70.3	71.7	71.9	67.8	60.8	50.3	33.4	15.1	4.3

SOURCES: For 1955–1975, Bureau of Labor Statistics (1980); for 1980, Bureau of Labor Statistics (1981:Annual Data Table 3); for 1985, Bureau of Labor Statistics (1986:Annual Data Table 3).

some difference in average turnover for women and men, women's labor force behavior is approaching that of men. Moreover, when the differences in earnings or occupations between women and men are taken into account, turnover differs very little between them (Reskin and Hartmann, 1986). Job tenure, a measure of attachment to a particular job rather than to the labor force as a whole, also shows convergence between women and men (Blau and Ferber, 1986:Table 7.8).

Another indicator of women's increased labor force participation and decreased turnover is their work-life expectancy. In 1979–1980 a 20-year-old woman could expect to work 27.2 years, compared with 14.5 years in 1950; a 20-year-old man could expect to work 36.8 years in 1979–1980 (S. Smith, 1985). The percentage of women working full time, year round has also increased steadily, although it is still less than that of men (Cain, 1985:Table 7). In 1983, 48 percent of women held full-time, year-round jobs, representing an increase of about 10 percentage points over a 20-year period. In 1983 the rate for men had not changed much and stood at about 64 percent (Blau and Ferber, 1986:75).

EDUCATIONAL ATTAINMENT OF THE LABOR FORCE

Education is an important factor influencing women's labor force participation and their earnings. Both the amount and type of education of workers are important for labor market outcomes. Educational preparation is especially important to workers' ability to adapt to change.

Throughout most of the history of public education in the United States, women have been about as well educated as men. Women have been more likely to complete high school than men, but men were introduced earlier and in larger numbers to college education. Thus, although the average years of schooling attained by women and men have been fairly equal, women and men have been distributed across levels of education differently. Men have had higher proportions of both high school dropouts and college degrees; women have had higher proportions of high school graduates. These trends are shown in Tables 3-5 and 3-6.

For the population as a whole, more men than women have completed college. In recent years women have earned almost as many college degrees as men (49.8 percent of four-year degrees in 1981; Blau and Ferber, 1986). Table 3-7 shows the changes in the population over age 25 who completed college between 1967 and 1981; it shows the effects of recent increases in college completion for women. It also shows increases in the proportion of black women and men who have completed college.

The educational attainment of men and women in the labor force has also been fairly equal. In 1962, median years of school completed were 12.3 for

TABLE 3-5 Mean Schooling Levels by Birth Cohort

Birth Cohort	Years of Schooling	
	Males	Females
1951–1954	12.57	12.65
1946–1950	12.62	12.39
1941–1945	12.21	12.05
1936–1940	11.85	11.70
1931–1935	11.50	11.39
1926–1930	11.17	11.16
1921–1925	10.89	10.97
1916–1920	10.46	10.56
1911–1915	9.83	10.09
1906–1910	9.41	9.75
1901–1905	8.85	9.15
1896–1900	8.44	8.71
1891–1895	7.92	8.19
1886–1890	7.51	7.89
1881–1885	7.31	7.70
1876–1880	7.20	7.65
1871–1875	6.92	7.27
1866–1870	6.79	7.16

SOURCE: Smith and Ward (1984:Table 17). Reprinted with permission.

TABLE 3-6 Schooling Distributions of Selected Birth Cohorts

Birth Cohort	Years of Schooling								
	0	Less Than 5	Less Than 9	8	Less Than 12	12	More Than 12	16	College Degree
Men									
1866–1870	7.5	21.1	77.8	35.5	85.8	9.0	5.2	1.6	1.9
1886–1890	5.7	17.4	66.3	28.4	79.3	12.0	8.8	2.3	3.3
1906–1910	1.4	6.6	41.6	20.3	61.5	23.1	15.4	4.6	6.3
1926–1930	0.7	2.8	17.4	8.2	38.6	42.3	19.1	5.7	7.7
1946–1950	0.4	1.5	6.6	2.3	19.1	45.6	35.3	12.9	15.9
Women									
1866–1870	8.5	26.0	82.1	33.6	88.1	5.7	6.2	2.2	3.3
1886–1890	6.4	21.6	71.9	28.2	82.1	8.6	9.0	2.8	4.6
1906–1910	1.8	8.8	47.5	22.3	67.0	18.0	15.0	4.7	7.8
1926–1930	1.0	3.9	22.8	10.5	43.5	31.2	25.3	8.7	14.2
1946–1950	0.5	1.2	7.8	3.0	19.4	37.6	43.1	17.6	20.4

SOURCE: Smith and Ward (1984:Table 18). Reprinted with permission.

TABLE 3-7 Population 25 Years or Older with Four or More Years of College, 1967–1981, by Gender, White, Black, and Spanish Origin

Year	Men	Women	White	Black	Spanish Origin[a]
			Percentage		
1967	12.8	7.6	10.6	3.9	n.a.
1968	13.3	8.0	11.1	4.3	n.a.
1969	13.5	8.2	11.2	4.6	n.a.
1970	14.1	8.2	11.8	4.5	n.a.
1971	14.6	8.5	12.0	4.5	n.a.
1972	15.4	9.1	12.6	5.1	n.a.
1973	15.9	9.6	13.1	6.0	n.a.
1974	16.9	10.1	14.0	5.6	n.a.
1975	17.6	10.6	14.6	6.5	8.4
1976	18.6	11.3	15.4	6.6	6.1
1977	19.2	12.1	16.2	7.3	6.2
1978	19.7	12.2	16.4	7.2	7.1
1979	20.4	12.9	17.2	7.7	6.6
1980	20.9	13.6	17.9	8.0	7.6
1981	21.1	13.4	17.8	8.2	7.7
			Number (thousands)		
1967	6,369	4,188	9,991	377	n.a.
1975	9,679	6,565	15,063	713	301
1981	13,208	9,466	20,775	1,084	485

[a] Persons of Spanish origin may be of any race.

SOURCES: Data for 1967 through 1974 from Bureau of the Census (1980:Table 6/15); data for 1975 through 1981 from Bureau of the Census, *Educational Attainment in the U.S.*, various years, all Table 1.

women and 12.1 for men; in 1983 they were 12.8 for both. Over those two decades the proportion of the labor force who were college graduates nearly doubled, from 9.7 to 18.4 percent for women, and from 11.9 to 23.1 percent for men. Table 3-8 shows the educational attainment of the labor force in 1983 by sex, race, and Spanish origin. Fewer white women had completed college than white men, and their median attainment was 0.1 year less than for men. Both black and Hispanic women had higher attainments than the men in their groups, and both were more likely to have completed high school or college or to have had some college education. Overall, however, the college attainment of minority workers in 1983 was similar to that for whites in the 1960s.

Despite the equality in years of schooling for men and women in the labor market, there are substantial differences in courses of study. In recent years there has been some convergence. Table 3-9 shows the change in the percentage of bachelor's degrees awarded to women in selected fields between 1966 and 1981. The percentage increased in every field except home economics, and at

TABLE 3-8 Educational Attainment of the Labor Force by Sex, Race, and Hispanic Origin, 1983[a]

Years of School Completed	Whites		Blacks		Hispanics	
	Males	Females	Males	Females	Males	Females
Fewer than 4 years of high school (%)	21.1	16.9	34.0	25.3	48.1	41.0
4 years of high school only (%)	37.6	44.9	39.8	43.5	30.4	35.9
1 to 3 years of college (%)	17.9	20.0	15.9	19.5	13.5	14.5
4 or more years of college (%)	23.4	18.2	10.3	11.7	8.0	8.6
Median school years completed	12.8	12.7	12.4	12.6	12.1	12.3

[a] Data are for workers aged 16 and over.

SOURCE: Blau and Ferber (1986:Table 7.2). Reprinted with permission.

least doubled in six fields: in agriculture 10 times, in architecture 4 times, in business 4 times, in computer and information science 2 times, in engineering 25 times, and in economics 3 times. Nevertheless, women are still substantially less likely than men to earn degrees in the physical sciences, engineering, or architecture. Although their distribution across fields has become more similar to men's, it still differs significantly. The index of segregation across 22 fields

TABLE 3-9 Percentage of Bachelor's Degrees Awarded to Women by Discipline, 1966 and 1981, Selected Fields

Discipline	1966	1981
Agriculture	2.7	30.8
Architecture	4.0	18.3
Biological sciences	28.2	44.1
Computer and information sciences	13.0[a]	32.5
Education	75.3	75.0
Engineering	0.4	10.3
English and English Literature	66.2	66.5
Foreign languages	70.7	75.6
Health	76.9	83.5
Home economics	97.5	95.0
Mathematics	33.3	42.8
Physical sciences	13.6	24.6
Psychology	41.0	65.0
Social sciences	35.0	44.2
Economics	9.8	30.5
History	34.6	37.9
Sociology	59.6	69.6
All fields	39.9[b]	49.8

[a] Data are for 1969, the earliest year available.
[b] Includes first professional degrees.

SOURCE: Blau and Ferber (1986:Tables 7.3 and 7.4). Reprinted with permission.

was 35.8 in 1981 (calculated from data in Women's Bureau, 1983:Table IV-13). This index can be interpreted as meaning that 35.8 percent of women or men would have to change fields for women and men to be identically distributed across fields.

The courses of study taken in high school also differ by sex. Among students enrolled in vocational courses, for example, girls are much more likely to take business courses (e.g., typing), and boys are much more likely to take craft and industrial arts courses (National Center for Education Statistics, 1984:Table 3.6). Equal proportions of high school boys and girls take first-year algebra, but girls are somewhat less likely to take higher-level math courses: for example, 10 percent of boys but only 6 percent of girls take precalculus or calculus (National Science Foundation, n.d.:Table II-9). And at two-year colleges (in 1978), male students outnumbered female students 10 to 1 in engineering, 3 to 1 in agriculture and natural resources, and almost 2 to 1 in physical sciences; women outnumbered men 12 to 11 in the biological sciences (National Science Foundation, n.d.:Table II-13).

The proportion of advanced degrees awarded to women has increased substantially since 1971. The proportion of first professional degrees awarded to women doubled between 1971 and 1975 and again to 1981, when women earned 27 percent of the degrees awarded (14.4 percent in dentistry, 24.7 in medicine, 42.6 in pharmacy, 35.2 in veterinary medicine, 32.4 in law, and 14.0 in the theological professions). With respect to other advanced degrees, women in 1981 earned 50.3 percent of the master's degrees awarded and 31.1 percent of the doctor's degrees.

In sum, although women and men have equal years of education on average, men are more likely to take mathematics, engineering, and physical science courses. Men are more likely to have advanced education. These differences are declining, however, as women's participation in college and graduate work has increased and their fields of study have changed. Overall, minorities have less education than whites, but this gap, too, has been declining.

THE POTENTIAL EFFECTS OF TECHNOLOGICAL CHANGE

THE INFLUENCE OF LABOR SUPPLY

The data reviewed above suggest that the labor force will continue to grow, although at slower rates than in the past. First, the population base from which the labor force will be drawn will experience slower growth. Most of the baby boom has entered the labor force, and the new replacement cohorts are smaller. Second, labor force participation rates for men are expected to continue their slow decline. Third, labor force participation rates for women, although they will continue to increase, are projected to do so at a slower rate. Although the rate of entry is expected to slow for women, women may be more likely to work full time. There will also be some shift in the composition of the available labor supply, with women becoming relatively more numerous in the labor force, and young people and older men becoming less numerous. Minorities are also expected to become a larger proportion of the labor force, since their birth rates are generally higher, and their current populations are younger. In addition, immigration is likely to continue to contribute disproportionately to the minority population and to result in the entry of predominantly working-age people.

These anticipated changes in the supply of different groups of workers to the labor market are likely to have effects on their average wage rates, because they have not in the past been seen as substitutes for one another. As noted in Chapter 1, there is substantial sex segregation in the labor market: many jobs are predominantly female, others predominantly male. There is also segregation by race and by age. Changes in the supply of women workers tend to affect women's wages relative to those of men because such changes tend to affect the supply to some occupations more than others. As the availability of the usual labor supply for a job changes, however, employers may change their hiring practices or make other adjustments in their production process.

The projected decreased rate of growth of female labor supply could be expected to contribute to a higher female-to-male wage ratio. In addition to the demographic reasons for a decreased supply of women workers noted above, changing attitudes and increased organization among women workers are likely to be important. Fewer women may be willing to work for low wages; they may demand full-time work, equitable wages, and seniority increases for their longer job tenure. Higher wages will certainly benefit women, but it is important to keep in mind that, in the long run, a higher price for female labor could in turn lead to substitution; younger people of both sexes, male immigrants or minorities, or automation might become feasible substitutes for adult, native-born, white women workers. Or, if the products that women workers produce become more expensive because of higher wages, consumers of those products might substitute cheaper products. Whether or how such substitution for women workers would occur is uncertain, but it is certainly one possible effect of supply changes that tend to raise the wages of women workers.

If economic growth remains at roughly current levels and widespread technological displacement does not occur, the employment prospects for adult women appear good. Some establishments that hire young workers are already complaining of shortages of labor. The shift in age and sex composition of the labor force suggests that additional shortages may be perceived before 1995, especially if economic growth increases. If economic growth slows or if technological change causes significant displacement of women workers from their jobs, however, serious unemployment for women may occur. Because of the variety of interrelated factors influencing women's labor force participation, women are unlikely to withdraw from the labor force simply because their jobs have been eliminated or new jobs fail to appear. Rather, a substantial number of women, like men, will most probably remain in the labor force as unemployed workers. Although "discouragement"—leaving the labor force because of lack of jobs, rather than lack of desire to work—is more pronounced among women, it is not likely to be large enough to alter this picture. For a smooth adjustment to large technological displacements, one would have to look to the job vacancies being created by the shrinking youth labor force, by the trend toward early retirement of older male workers, and by the increased availability of "men's" jobs in response to equal employment opportunity programs. Or one must hope that the change will occur slowly enough that it can be relatively easily accommodated.

A well-educated work force is clearly better able to respond to occupational shifts in the demand for labor. People who have completed a solid curriculum in high school will be able to learn new skills on the job and benefit from retraining (see Committee on Science, Engineering, and Public Policy, 1984). Those with scientific and technical training will also be able to enter rapidly expanding fields with such jobs as computer systems analysts, programmers, and operators; electrical engineers; and electrical and electronics technicians. These jobs, although not among the largest occupations, are among the top 40 occupations in the number of new jobs they are expected to create by 1995 (Committee on Science, Engineering, and Public Policy, 1984). To the extent that workers lack basic educational preparation and technical and scientific background, they are likely to be less able to respond to forthcoming changes. As noted above, in some aspects women and minorities lag behind majority men in their educational preparation.

THE DEMAND FOR WORKERS

As noted in Chapter 1, technological change can affect the demand for workers in several ways.

First, new technology may cause an increase in labor productivity, making it possible to produce the same output with a smaller number of workers. As a consequence, fewer workers may be demanded than previously. As noted in Chapter 1, technological change permitted the nation's food supply to be produced by an ever-shrinking proportion of the work force. Consumers and the economy as a whole benefited. A large proportion of the work force left food production, while employment in manufacturing and service jobs increased. Economic output grew and per capita income increased. For individuals caught up in this transition, however, the costs of adjustment were often high.

A decline in employment because of increased labor productivity is not inevitable, however. By facilitating such increases, new technologies decrease the cost of producing the product and thus lower its price. The lower price in turn may encourage consumers to demand more of the product. In addition, technological change may allow for the provision of a higher-quality product or it may stimulate the introduction of new products, increasing consumer demand for the output of the industry. In the case of food production, the increase in labor productivity was very large relative to the possible increase in demand for food by consumers. Furthermore, the new products that resulted—canned, frozen, processed, and packaged foods—were not produced by farm workers. Thus, an enormous decline in the proportion (as well as the absolute number) of farm workers occurred. In other industries, aggregate demand and newly created products and services are likely to be more favorable to employment levels.

As an example, the adoption of word-processing equipment may indeed increase labor productivity, permitting the existing work to be accomplished by fewer workers. But its introduction is also likely to create new work. The capabilities of microprocessors are ideally suited to many revisions, more personalized form letters, updated statistical reports, and more charts and graphs, all of which can be produced more cheaply than previously. As a consequence, the production of a firm becomes more intensive in information content, a trend that has been going on for many years, and the decline in employment, to the extent it occurs, is considerably less than if the new technology were used simply to produce the previous output. Thus, while a major impetus behind the adoption of new technologies is the increase in labor productivity that they permit, their impact on employment levels is difficult to forecast.

A second potential impact of the application of new technologies is that they can facilitate the reallocation of functions among occupations, resulting in either a reduction or an increase in the demand for a particular type of worker. For example, the greater ease of data entry and retrieval using the new computer technologies may encourage employers to shift some of these tasks from clerical workers to other workers or even to consumers (as in banking). The many capabilities and ease of use of the latest word-processing equipment and software may also result in professional and managerial workers doing more of their own "typing." Such developments would reduce the demand for clerical workers. In the opposite direction, however, the application of computers can make routine some decisions that formerly required broad-based knowledge and judgment, so functions previously performed by managerial or professional workers would be transferred to clerical workers, increasing the demand for their services. For example, in the insurance industry, rating and underwriting are now often done by computers using standardized decision algorithms: a clerical worker can enter the data, describing the client and his or her needs, and, if there are no special conditions, issue a policy.

A third impact of the introduction of new technologies is its tendency to alter the skills needed to produce the former (or new) output. It may render specific skills obsolete, or nearly so, and greatly increase the demand for others. Such effects are evident when looking at specific occupational categories, such as stenographers and computer operators. Between 1972 and 1982, when clerical employment overall rose by 28.8 percent, employment of stenographers declined by 47.2 percent (Hunt and Hunt, 1985a). This decrease was caused by the miniaturization of dictation equipment and improvements in magnetic tape technology combined with growing consumer acceptance of new methods. When dictation equipment is used, the material is entered by the clerical workers directly into the typewriter or word processor, and the shorthand skills of the stenographer are no longer needed. The decline in the number of stenographers is not a reflection of a decline in the amount of dictation being done—indeed, the amount of dictation appears to be increasing—but the particular skills of stenographers are no longer needed to accomplish this task. Over the same 10-year period, the employment of computer and peripheral equipment operators increased 195.5 percent. This increase was clearly fueled by the diffusion of the new computer technologies. It should be noted, however, that changes in skills needed to accomplish specific tasks do not always translate into changes in occupational employment, because single tasks do not often predominate in jobs. Thus, rather than reducing employment of secretaries, automated word processing may simply provide them with more time for other work.

In sum, the introduction of new technologies may result in increases in the productivity of affected workers, reallocation of functions among occupations, and changes in the skill requirements of jobs. For the reasons outlined above, it is difficult to predict the consequences of these processes on employment levels of workers in any given occupational category. Moreover, the factors may operate together to make such forecasts even more uncertain. So, for example, with word-processing technology, it may be possible to generate frequent statistical reports by simply entering a few updated numbers. While some new skills will need to be learned, others will not be needed, and, overall, the greater ease of production may create a demand for additional statistical reports. But it is not certain that those reports will be generated by clerical workers.

It is useful to keep in mind that these effects of technological change on the demand for workers in different occupations may affect the wage rates of workers as well as their employment levels. In the long term, changes in wage rates can influence employment levels. At issue here is the fact that different effects on different occupations are likely to affect women and men differently, changing the female-to-male wage ratio.

Unemployment

Of the three major types of unemployment—"frictional" (the expected unemployment that would normally result from workers' job changes and new entrants), "structural" (the unemployment that remains at the peak of the economy's upswings, and is not frictional, being caused by changes in the structure of demand), and "cyclical" (which comes and goes with the ebb and flow of the business cycle)—structural unemployment in particular has troubled observers since the late 1960s.

Sources of structural unemployment, caused by imbalance between the types and locations of available employment on the one hand and the qualifications and locations of workers on the other hand, include skill mismatch, geographic mobility and immobility, discrimination, and institutional barriers to employment. When structural factors exist, both unemployment and shortages occur, despite a balance between supply and demand in the economy as a whole. Technological change probably contributes most to the first factor, but may contribute as well to the importance of the second if it increases the mobility of capital. It is not, in any case, the sole cause of structural unemployment.

"Displaced workers" are those who are laid off more or less permanently; their jobs may have disappeared because of technological change, competition, or decreased demand for the products they make. Displaced workers and new entrants to the labor market may or may not be considered structurally unemployed, depending on how quickly and easily they can find new employment.

Because the adoption of new technologies almost invariably results in shifts in demands for different types of workers, some displacement of workers is likely to occur. Even if, due to counterbalancing effects, broad categories are relatively unaffected, specific occupational groups are certain to be affected. The consequences of such shifts in occupational demand for unemployment among workers in the affected groups depend primarily on the magnitude of the change, the speed with which it occurs, and the flexibility and skills of the workers involved. Relatively small changes in demand or those that take place slowly are likely to cause little unemployment. At the level of firms, normal attrition may be sufficient to reduce the size of the work force in the affected occupations. At the level of the labor force, attrition may also operate as older workers retire from declining occupations and are not replaced by new entrants.

Relatively large changes, however, especially when they occur over a relatively short period of time, can cause more serious problems. Workers may lose their jobs and find themselves without the skills necessary to obtain new jobs. Thus, the adoption of new technologies has the potential to create groups of workers who are structurally unemployed. The magnitude of this problem depends not only on the extent and timing of shifts in demand but also on the resources devoted by firms, as well as the public sector, to retraining and placing those workers. Workers can be encouraged to move where there are jobs; they can be reeducated, retrained, or retired; capital movement can be controlled; discrimination can be reduced; and institutions can be modified.

Increased structural unemployment, should it occur, can make it difficult for the economy to achieve desired low rates of unemployment at acceptable rates of inflation: if there is a growing pool of workers who lack the necessary skills for the available jobs, increases in aggregate demand will rapidly generate shortages of qualified workers. As the wages of those workers are bid up, labor costs and thus prices rise. This phenomenon may be an important factor in the rising trend of unemployment over the past two decades. The policy priority of the government has been on reducing inflation, which, given larger structural shifts, causes unemployment to increase. If structural unemployment increases, a further upward revision in the government's target rate of unemployment is likely to occur.

Several observers have noted that, since 1969, the economy has experienced increased unemployment (e.g., Podgursky, 1984). At the peak and the trough of each of the four business cycles between 1969 and 1982, unemployment has been greater than before. At the peak of the first cycle (in the third quarter of 1969), the unemployment rate was 3.6 percent; at the peak of the fourth cycle (in the third quarter of 1981), it was 7.4 percent. Similarly, at the trough of the first cycle (in the fourth quarter of 1970), unemployment was 5.8 percent; by the trough of the fourth cycle (in the fourth quarter of 1982), it was 10.6 percent. In his analysis, Podgursky (1984) finds that the large number of young people and women who entered the labor market in the early years (early 1970s) contributed to the increase, and that in later years (late 1970s, early 1980s), male unemployment in blue-collar jobs contributed disproportionately to the increased unemployment. Of special interest is that in the most recent trough the share of unemployment attributable to clerical occupations grew. Most importantly, Podgursky finds that the secular increase in total unemployment almost certainly points to rising structural unemployment and a probable difficulty in absorbing the employment effects of technological change. Although he notes that the evidence also suggests there has been some slackening in the strength of aggregate demand over time, the increased structural unemployment suggests a need for specifically targeted employment programs; changes in monetary and fiscal policies that affect aggregate demand are not likely to reach all the structurally unemployed. There is nothing immutable about the residual unemployment left at the peak of recovery or about the increasingly high levels of unemployment at the trough of the cycle. Both are susceptible to structural solutions, and when aggregate demand is strong enough (as in wartime), structural solutions become worthwhile (for example, child care programs in World War II; training programs during the Vietnam War).

Important current structural factors include skill mismatches between the available workers and the available jobs, geographic imbalances (workers seeking jobs do not live where jobs are available), differences in expectations between workers and employers (unemployed workers do not want the available jobs because of their working conditions; employers do not want to hire available workers because they believe them to be unqualified or because they are prejudiced), and other barriers to entry or movement. From a structural viewpoint, the displacement of women may be especially serious because women workers have not often been viewed by employers as interchangeable with men and because they tend to be less mobile geographically and occupationally.

Although it has not been established that the new computer-based technologies are associated with this worsening problem of structural unemployment, the increased contribution of clerical work to unemployment in the last cycle raises the possibility that they might be involved. Moreover, information technologies can be expected to have greater effect in the future as their use becomes more widespread. As noted earlier, the predominantly female clerical occupations are likely to be the most affected.

RECENT TRENDS IN CLERICAL EMPLOYMENT

OVERALL GROWTH

While forecasting inevitably holds many uncertainties, the best starting point for trying to predict future clerical employment is to review recent trends. This section and the next one on future employment are based largely on work done for the panel by H. Allan Hunt and Timothy L. Hunt (1985a).[2] The tremendous historical growth in clerical employment is illustrated in Figure 3-1, which shows that the proportion of clerical workers to total employment has *doubled* in the last 40 years. In 1940 just under 1 employee in 10 was a clerical worker; by 1980 this proportion had risen to nearly 1 in 5; the number of clerical workers increased from 5 million to 20 million (Hunt and Hunt, 1985a).

Will this trend continue? Both the increasing unemployment evident in the past four business cycles and the rapid pace of the technical developments in the office described in Chapters 1 and 2 cast doubt on future growth. The first "computer revolution" in the 1960s was also expected to decrease the need for clerical workers, but clerical employment continued to increase. Nevertheless, many observers are convinced that the new technologies will substantially reduce clerical employment. Reductions in the cost of computing hardware, combined with the reductions in size made possible by microprocessor technology, facilitate the widespread use of the new equipment in many settings. These changes may actually constitute a revolutionary development.

As Hunt and Hunt (1985a) note, some support for the position that the new computer revolution will stem the rate of growth in clerical employment is

---

[2] The data developed by Hunt and Hunt have been adjusted rather extensively for consistency (see Hunt and Hunt, 1985a, for a detailed explanation of the adjustments). Thus, the figures reported here do not correspond exactly with published Census Bureau figures.

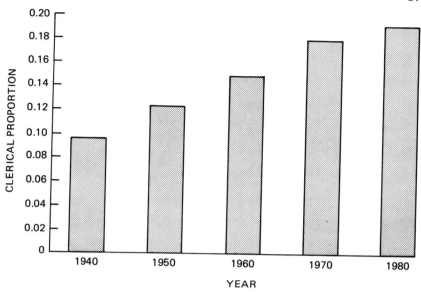

FIGURE 3-1 Decadal growth in clerical employment as a proportion of total employment, 1940–1980. SOURCE: Hunt and Hunt (1985a:Figure 1-A); based on 1940–1980 census data.

found in the apparent reduction in the rate of increase of clerical workers as a proportion of the total labor force. While the clerical proportion rose almost linearly from 1940 to 1970, there is a slight reduction in the rate of increase between 1970 and 1980 (see Figure 3-1). Is this the beginning of the end of the growth of clerical employment? The question cannot yet be answered. In addition to the relatively small number of observations for the post-1980 period, the data for those years are influenced by the major recession of 1981 to 1982, the deepest in terms of unemployment since the 1930s, and the recovery from it, making it hard to separate short-run cyclical influences from long-run trends. In past recessions the proportion of clerical employment tended to increase since employers usually cut back less on clerical workers than on other categories of employees. Significantly, this effect did not hold in the 1981–1982 recession. Overall, as may be seen in Figure 3-2, clerical employment as a proportion of total employment actually declined slightly in the early 1980s, following the above-noted reduction in the rate of increase in the proportion of clerical employment during the 1970s. While the available evidence is consistent with the hypothesis of slowed growth in clerical employment, however, there is little in past trends that would support a projection of catastrophic decline in this occupational group in the near future.

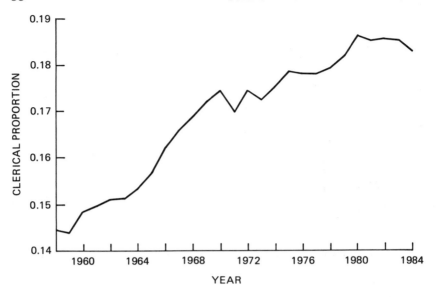

FIGURE 3-2 Annual changes in clerical employment as a proportion of total employment, 1958–1984. SOURCE: Hunt and Hunt (1985a:Figure 1-C); based on 1958–1984 CPS data.

OCCUPATIONAL SHIFTS WITHIN CLERICAL WORK

Within the clerical category there has been considerable variation over the past 30 years in how specific occupations have fared. Table 3-10 ranks clerical occupations by the rate of change in employment levels from 1950 to 1980. The rate of increase in this period was by far the largest for computer and peripheral equipment operators, although this occupation grew from a very small base, 868 people in 1950. By 1980 there were nearly 400,000 computer and peripheral equipment operators, making it the thirteenth largest clerical occupation. Teachers' aides were the second fastest growing clerical occupation over the 30-year period; their numbers also grew from a fairly small base, 6,000 in 1950, to more than 200,000 by 1980. The third fastest growing clerical occupation was typists, although there was actually a significant decline from 1970 to 1980; the rapid growth of the occupation from 1950 to 1970 offset the recent decline. Following in order of rate of growth were library attendants, clerical supervisors, bank tellers, receptionists, and cashiers.

With the exception of the computer operator category, the jobs that grew rapidly do not appear to be related to the use of new technologies. It is also important to recognize that very large categories such as secretary provided a high proportion of new jobs over the period, even at lower overall rates of growth.

A few clerical occupations showed declines between 1950 and 1980. The most rapid declines were among stenographers and telegraph operators; both of these categories were affected by technology. Demand for telegraphic services has fallen off as cheaper, more convenient long distance communication has become available. As discussed above, the use of dictation equipment has contributed to the decline of stenographers. Yet the functions of long distance communication and of dictation and transcription are, if anything, increasing, assisted by new equipment and different kinds of workers.

Tabulating machine operators and weighers also declined substantially. As Hunt and Hunt (1985a) note, tabulating machine operators illustrate how a technology-specific occupation can experience a spurt of rapid growth and then an equally sudden decline. Tabulating machines were developed in the 1950s for analyzing data on punched cards, and their use grew rapidly. Between 1950 and 1960, the number of tabulating machine operators nearly tripled. But the adoption of new technology with more advanced capabilities reduced the use of tabulating machines, and the number of employees fell by nearly 90 percent after 1960. Other declining occupations between 1950 and 1980 were those of messengers and office helpers, calculating machine operators, and telephone operators. These declines appear to be related to the availability and use of new technologies.

Demographic Trends in Clerical Employment

The differential rates of change among the different clerical occupations have affected women and men and minority and majority women somewhat differently because of their different employment patterns in those occupations. The fastest-growing occupations between 1950 and 1980 (see Table 3-10) are generally among the most predominantly female of the clerical occupations: typists, receptionists, teachers' aides, bank tellers, and keypunch operators. One of the rapidly growing occupations over the period—insurance adjusters, examiners, and investigators—although still more male (44 percent in 1980) than clerical occupations as a whole (22 percent), has experienced rapid increases in the percent female: overall employment in the occupation grew at a rate of 5.4 percent per year, while female employment grew 16.1 percent per year (Hunt and Hunt 1985a:Tables 2.9, 2.15). In contrast to these rapidly growing occupations, many of the clerical occupations in which males are relatively overrepresented have experienced average or below-average growth: utility meter readers; messengers and office helpers; shipping and receiving clerks; ticket, station, and express agents; and postal office mail carriers. Indeed, over the period as a whole, females in all clerical occupations increased from 61 to 78 percent.

Minority representation in clerical occupations also increased over the period: CPS data from 1972 to 1982 show an increase from 8.7 to 11.8 percent

TABLE 3-10 Employment in Clerical Occupations, 1950 to 1980, Ranked by Relative Change 1950 to 1980

Occupational Title	Employment 1950	1960	1970	1980	Annual Percent Change
Computer and peripheral equipment operators	868	2,023	124,684	391,909	22.6
Teachers' aides, except school monitors	6,105	17,804	139,790	207,391	12.5
Typists	60,534	547,923	1,041,804	799,561	9.0
Library attendants and assistants	16,235	38,203	133,911	140,808	7.5
Clerical supervisors, n.e.c.	44,348	56,887	119,887	340,946	7.0
Bank tellers	66,944	139,477	265,197	476,233	6.8
Receptionists	77,965	164,446	323,552	536,963	6.6
Cashiers	252,252	510,179	884,531	1,654,151	6.5
Office machine operators	146,778	326,521	588,356	890,288	6.2
Keypunch operators	75,091	169,000	290,119	382,118	5.6
Insurance adjusters, examiners, and investigators	33,061	58,726	102,043	159,124	5.4
Miscellaneous clerical workers	253,633	328,399	506,677	1,163,635	5.2
Counter clerks, except food	96,313	127,630	243,697	398,029	4.8
Office machine, n.e.c.	9,788	21,352	38,669	39,864	4.8
Secretaries	1,005,968	1,539,017	2,875,826	4,058,182	4.8
Estimators and investigators, n.e.c.	112,469	171,901	282,074	442,553	4.7
Billing clerks	32,357	45,254	112,876	117,943	4.4
Real estate appraisers	11,754	15,822	22,735	41,343	4.3
Mail handlers, except post office	53,563	67,300	133,839	182,223	4.2
Payroll and timekeeping clerks	65,697	112,901	165,815	218,387	4.1
Duplicating machine operator	5,520	14,392	21,682	17,971	4.0

Collectors, bill and account	25,395	34,229	54,728	76,982	3.8
Statistical clerks	109,956	143,922	265,431	297,939	3.4
File clerks	118,211	152,160	382,578	316,419	3.3
Expediters and production controllers	123,277	151,191	217,107	329,621	3.3
Dispatchers and starters, vehicle	33,746	49,205	63,699	87,622	3.2
Bookkeepers	744,053	973,224	1,633,490	1,804,374	3.0
Ticket, station, and express agents	69,807	76,994	104,285	152,841	2.6
Proofreaders	12,708	17,171	29,940	27,321	2.6
Stock clerks and storekeepers	274,089	384,115	482,259	580,979	2.5
Not specified clerical workers	1,185,906	1,610,020	862,394	1,880,102	1.5
Mail carriers, post office	164,851	203,116	268,612	258,966	1.5
Shipping and receiving clerks	323,785	325,307	100,890	483,183	1.3
Postal clerks	216,164	242,872	321,263	315,111	1.3
Bookkeeping and billing machine operators	26,610	53,914	67,341	37,200	1.1
Enumerators and interviewers	85,013	118,723	68,697	88,712	0.1
Meter readers, utility	40,696	39,712	35,144	41,407	0.1
Calculating machine	19,176	38,903	37,153	17,881	−0.2
Telephone operators	363,472	374,495	433,739	314,674	−0.5
Messengers and office helpers	111,508	61,303	61,050	82,225	−1.0
Weighers	80,915	44,548	41,410	29,717	−3.3
Tabulating machine operator	9,725	26,937	8,685	3,345	−3.5
Telegraph operators	34,811	21,064	13,052	7,604	−4.9
Stenographers	429,424	283,486	136,197	91,593	−5.0

NOTE: n.e.c., not elsewhere classified.

SOURCE: Hunt and Hunt (1985a:Table 2.9); based on decennial census data, adjusted for consistency by the authors.

(Hunt and Hunt, 1985a:Table 2.17). Some of those occupations with relatively greater minority representation grew relatively rapidly: for example, teachers' aides (20.1 percent minority in 1982), typists (17.4 percent), keypunch operators (20.3 percent), and computer and peripheral equipment operators (15.5 percent). Other occupations, however, experienced very slow growth: stock clerks and storekeepers (16.1 percent minority), telephone operators (17.3 percent), messengers and office helpers (19.1 percent), and postal clerks (26.9 percent). Table 3-11 displays the numbers and percentages of several minority groups (blacks, Hispanics, and Asians) among female clerical workers in 1980 for detailed census occupational categories. Asian women, for example, are 1.6 percent of the female clerical work force, and they vary from 3.1 percent of transport, ticket, and reservation agents to 0.5 percent of female meter readers. Hispanic women, 4.6 percent of female clerical employment overall, range from 11.4 percent of female teachers' aides to 2.6 percent of female proofreaders. Black women, who are 9.2 percent of female clerical employment, are 32.9 percent of female postal clerks, but only 4.0 percent of bookkeepers and accounting clerks.

Clerical employment growth has also varied by region and geographic area. Although a thorough analysis of such differences is beyond the scope of this report, several recent studies highlight geographic changes that are significant for women and minorities. For example, a recent study by 9-to-5 (1985) found that, like blue-collar employment, clerical employment in the Midwest has declined since 1980, associated with the recessions in 1980 and 1982. Unlike the nation, this region experienced actual declines, not only lack of growth. Surprisingly, clerical employment fell even more rapidly than total employment between 1980 and 1983 in the four states (Illinois, Indiana, Michigan, and Ohio) that constitute the most industrial portion of the region. This study also found that the decline in clerical employment was especially large in center cities, such as Detroit and Chicago, while employment in the suburban areas of those cities either did not decline as much or expanded.

The 9-to-5 report traces the declines in clerical employment in the Midwest generally to declines in government employment. A recent study by the New York regional office of the Bureau of Labor Statistics found that finance job growth had fallen off steeply between 1983 and 1984 in Manhattan: in 1983 the finance industries had added 10,900 jobs (and 16,700 in 1982), but in 1984 only 2,300 new finance jobs were created (*New York Times,* December 15, 1985:59). Technological change and the movement of jobs to the suburbs were cited as causes for the slower growth. In research on insurance companies in the San Francisco Bay area, Nelson (1983) found similar suburbanization of jobs occurring as back offices were relocated from center cities to suburbs. She concluded that land cost was only one factor in the decision to relocate; management also sought a different labor force than that available in the cities. The

TABLE 3-11 Employment of Females in Administrative Support Occupations by Race and Spanish Origin in 1980

Occupation	Total	White Number	White Percent	Black Number	Black Percent	Hispanic Number	Hispanic Percent	Asian Number	Asian Percent
Administrative support occupations	12,997,076	11,325,716	87.1	1,200,516	9.2	595,461	4.6	205,036	1.6
Supervisors, administrative support occupations	497,668	430,522	86.5	51,481	10.3	20,578	4.1	7,160	1.4
Supervisors, general office	354,410	304,707	86.0	38,027	10.7	15,495	4.4	4,955	1.4
Supervisors, computer equipment operators	12,392	10,683	86.2	1,248	10.1	530	4.3	253	2.0
Supervisors, financial records processors	77,172	70,042	90.8	4,868	6.3	2,608	3.4	1,462	1.9
Chief communications operators	22,898	19,261	84.1	3,243	14.2	714	3.1	141	0.6
Supervisors, distributing, scheduling, and adjustment clerks	30,796	25,829	83.9	4,095	13.3	1,231	4.0	349	1.1
Computer equipment operators	241,155	202,392	83.9	28,374	11.8	11,396	4.7	5,272	2.2
Computer operators	226,354	189,858	83.9	26,597	11.8	10,666	4.7	4,980	2.2
Peripheral equipment operators	14,801	12,534	84.7	1,777	12.0	730	4.9	292	2.0
Secretaries, stenographers, and typists	4,579,938	4,115,730	89.9	325,679	7.1	186,010	4.1	55,143	1.2
Secretaries	3,823,248	3,507,644	91.7	214,543	5.6	142,531	3.7	39,681	1.0
Stenographers	77,841	67,511	86.7	7,416	9.5	2,871	3.7	1,643	2.1
Typists	678,849	540,575	79.6	103,720	15.3	40,608	6.0	13,819	2.0
Information clerks	763,561	667,984	87.5	65,629	8.6	41,366	5.4	10,901	1.4
Interviewers	104,582	89,229	85.3	10,702	10.2	5,223	5.0	1,572	1.5
Hotel clerks	41,756	37,746	90.4	2,368	5.7	1,620	3.9	897	2.1
Transportation, ticket, and reservation agents	57,161	47,982	83.9	6,418	11.2	3,419	6.0	1,785	3.1
Receptionists	494,800	438,214	88.6	37,819	7.6	27,988	5.7	5,865	1.2
Information clerks, n.e.c.	65,262	54,813	84.0	8,322	12.8	3,116	4.8	782	1.2
Nonfinancial records processing	745,372	616,397	82.7	96,966	13.0	40,658	5.5	13,579	1.8
Classified-ad clerks	10,521	9,478	90.1	695	6.6	386	3.7	148	1.4
Correspondence clerks	15,741	13,031	82.8	2,321	14.7	552	3.5	152	1.0
Order clerks	209,871	177,969	84.8	24,654	11.7	11,615	5.5	2,670	1.3

Continued

TABLE 3-11 (Continued)

Occupation	Total	White		Black		Hispanic		Asian	
		Number	Percent	Number	Percent	Number	Percent	Number	Percent
Personnel clerks	65,759	54,709	83.2	7,862	12.0	3,490	5.3	1,473	2.2
Library clerks	114,294	97,642	85.4	11,741	10.3	4,324	3.8	2,765	2.4
File clerks	221,350	171,860	77.6	37,927	17.1	14,917	6.7	4,370	2.0
Records clerks	107,836	91,708	85.0	11,766	10.9	5,374	5.0	2,001	1.9
Financial records processing	1,991,619	1,834,478	92.1	96,239	4.8	67,451	3.4	32,936	1.7
Bookkeepers and accounting clerks	1,640,233	1,526,300	93.1	66,061	4.0	52,075	3.2	26,479	1.6
Payroll clerks	132,622	117,420	88.5	10,670	8.0	5,271	4.0	2,002	1.5
Billing clerks	115,020	102,349	89.0	8,772	7.6	5,090	4.4	1,869	1.6
Cost and rate clerks	58,731	51,014	86.9	5,372	9.1	2,481	4.2	1,328	2.3
Billing, posting, calculating machine operators	45,013	37,395	83.1	5,364	11.9	2,534	5.6	1,258	2.8
Duplicating, mail, office machine operators	38,462	30,359	78.9	6,272	16.3	2,170	5.6	762	2.0
Duplicating machine operators	11,484	8,998	78.4	1,931	16.8	521	4.5	275	2.4
Mail and paper handling machine operators	4,390	3,686	84.0	531	12.1	247	5.6	52	1.2
Office machine operators, n.e.c.	22,588	17,675	78.2	3,810	16.9	1,402	6.2	435	1.9
Communications equipment operators	276,148	228,320	82.7	38,887	14.1	12,493	4.5	2,602	0.9
Telephone operators	265,938	219,230	82.4	38,033	14.3	12,189	4.6	2,495	0.9
Telegraphers	2,711	2,241	82.7	357	13.2	85	3.1	59	2.2
Communications equipment operators, n.e.c.	7,499	6,849	91.3	497	6.6	219	2.9	48	0.6
Mail and message distribution clerks	229,096	169,374	73.9	51,173	22.3	9,136	4.0	3,617	1.6
Postal clerks	95,511	60,611	63.5	31,459	32.9	3,092	3.2	1,734	1.8
Mail carriers, postal service	33,179	28,638	86.3	3,738	11.3	928	2.8	268	0.8
Other mail clerks	79,425	62,818	79.1	13,035	16.4	4,070	5.1	1,340	1.7
Messengers	20,981	17,307	82.5	2,941	14.0	1,046	5.0	275	1.3

Material recording, scheduling, and distributing	571,300	485,198	84.9	64,039	11.2	30,734	5.4	8,493	1.5
Dispatchers	29,568	25,715	87.0	2,908	9.8	1,188	4.0	321	1.1
Production coordinators	112,539	97,423	86.6	11,150	9.9	5,569	4.9	1,672	1.5
Traffic, shipping, and receiving clerks	113,554	96,751	85.2	12,130	10.7	7,455	6.6	1,429	1.3
Stock and inventory clerks	198,345	167,381	84.4	23,021	11.6	9,858	5.0	3,479	1.8
Meter readers	4,239	3,546	83.7	553	13.0	193	4.6	21	0.5
Weighers, measurers, and checkers	26,348	21,441	81.4	3,791	14.4	1,769	6.7	233	0.9
Samplers	1,157	1,026	88.7	110	9.5	48	4.1	6	0.5
Expediters	57,242	49,058	85.7	6,361	11.1	2,466	4.3	761	1.3
Material recording, n.e.c.	28,308	22,857	80.7	4,015	14.2	2,188	7.7	571	2.0
Adjusters and investigators	321,234	265,850	82.8	42,185	13.1	15,844	4.9	6,334	2.0
Insurance adjusters, examiners, investigators	98,407	81,382	82.7	13,231	13.4	3,596	3.7	2,283	2.3
Noninsurance investigators and examiners	151,951	126,338	83.1	19,665	12.9	7,358	4.8	2,866	1.9
Eligibility clerks, social welfare	19,744	14,565	73.8	3,831	19.4	1,918	9.7	548	2.8
Bill and account collectors	51,132	43,565	85.2	5,458	10.7	2,972	5.8	637	1.2
Miscellaneous administrative support occupations	2,741,523	2,279,112	83.1	333,592	12.2	157,625	5.7	58,237	2.1
General office clerks	1,353,251	1,125,866	83.2	167,590	12.4	74,813	5.5	27,181	2.0
Bank tellers	451,465	401,512	88.9	32,136	7.1	22,163	4.9	8,573	1.9
Proofreaders	21,610	19,447	90.0	1,526	7.1	567	2.6	294	1.4
Data-entry keyers	349,477	269,598	77.1	57,126	16.3	21,516	6.2	13,106	3.8
Statistical clerks	104,345	86,019	82.4	14,099	13.5	4,624	4.4	1,871	1.8
Teachers' aides	191,564	147,189	76.8	31,569	16.5	21,780	11.4	2,125	1.1
Administrative support, n.e.c.	269,811	229,481	85.1	29,546	11.0	12,162	4.5	5,087	1.9

NOTE: n.e.c., not elsewhere classified.

SOURCE: Hunt and Hunt (1985a:Table 2.6); based on 1980 decennial census data.

suburban labor force is more likely to be white, college-educated, and married, and more willing to work part time. Shifts such as these in San Francisco, New York, and the Midwest clearly have negative implications for minority women in clerical jobs.

Sources of Change in Clerical Work

Sources of change in the size of occupations can be understood as resulting from three factors: overall economic growth; the differential growth rates of various industries, each of which uses occupations in different proportions (for example, finance uses relatively more clerical workers than manufacturing, while manufacturing uses more assemblers than finance); and the changes in the intensity with which the various industries use each occupation (for example, the insurance industry may alter its use of sales personnel relative to other workers over time). This last "intensity" factor can be measured by an occupational staffing ratio, which indicates the importance of a single occupation relative to total employment in the industry (see Hunt and Hunt, 1985a:Chapter 4, for a more complete discussion).

Industries vary in their relative importance in the economy, and, as noted, different industries use different mixes of occupations to produce their final output. Table 3-12 summarizes some of this basic information for 11 broad industrial categories in 1982, with particular emphasis on clerical employment. As shown, clerical staffing ratios vary from only 2.4 percent in agriculture to 43.9 percent in finance. Since the number of clerical jobs in a given industry is the product of the total employment level in the industry and the staffing ratio for clerical workers in that industry, an industry may employ a large number of clerical workers even though it has a relatively low staffing ratio for clerical workers if its total employment is large.

For example, as Table 3-12 shows, the service industry group employs about 5.5 million clerical workers, more than any other industry, although its clerical staffing ratio is not especially high. Slightly fewer than 3 million clerical jobs are located in each of the next two largest employers of clerical workers—retail trade and finance. Although clerical workers are dispersed broadly throughout the economy, these three industries combined—services, retail trade, and finance—account for more than 11 million clerical jobs, almost 60 percent of total clerical employment.

In terms of detailed industries, the 20 largest employers of clerical workers in 1982 are shown in Table 3-13, ranked by the total clerical employment in the industry.[3] Thus, the first industry listed, state and local government and educa-

---

[3] The total number of 105 industries in this analysis represents the entire economy. Hunt and Hunt (1985a) found they could adjust the industrial categories of both the CPS data and the Bureau of Labor Statistics OES data into 105 compatible categories.

TABLE 3-12 Employment by Industry, 1982

Industry	Total Industry Employment (thousands)	Percent of Total Employment	Percent of Workers in Industry in Clerical Jobs[a]	Clerical Employment (thousands)	Percent of All Clerical Workers Employed in Industry
Agriculture	3,401	3.4	2.4	83	0.4
Mining	1,028	1.0	12.5	128	0.7
Construction	5,756	5.8	7.8	451	2.4
Durables	11,968	12.0	12.6	1,513	8.2
Nondurables	8,318	8.4	12.9	1,074	5.8
Utilities	6,552	6.6	22.3	1,463	7.9
Wholesale trade	4,120	4.1	20.5	844	4.6
Retail trade	16,638	16.7	17.1	2,840	15.4
Finance	6,270	6.3	43.9	2,750	14.9
Services	30,259	30.4	18.1	5,473	29.7
Public administration	5,218	5.2	35.0	1,827	9.9
Total	99,528	100.0	18.5	18,466	100.0

[a]This percentage is also known as the clerical staffing ratio.

SOURCE: Calculated from Hunt and Hunt (1985a).

tional services, has the largest number of clerical employees, 2.5 million, and, as the fourth column shows, accounts for more than 13 percent of all clerical employment. The table also shows clerical staffing ratios and total industry employment. In addition to state and local government and education, the largest employers of clerical workers are the federal government, banking, insurance, communications and transportation industries, and a variety of service sector industries from business services to personal services. The top 10 employers of clerical workers account for about two-thirds of all clerical employment, while the top 20 industries account for more than 80 percent.

Since industry employment is crucial to assessing occupational employment trends, the trends over the last 27 years in total industry employment are presented in Figure 3-3, which aggregates the employment in the top 10 clerical-employing industries and shows it relative to employment in the 105 industries that constitute the total economy in this analysis. The numbers are reported in index number form to make it easier to compare the growth trends in the top 10 with average growth in all 105 industries. As can be seen in Figure 3-3, the top 10 clerical-employing industries show much less cyclical variation in employment than the economy as a whole. Employment in these 10 industries taken together remained positive through two of the three recessions during the period; only in 1982 did the composite growth rate of these 10 industries become negative, and then only barely so. Also important is that the average growth rate of these 10 industries clearly outdistanced the all-industry average for the entire 27-year period. As Hunt and Hunt (1985a) note, their higher growth stems

TABLE 3-13 Clerical Employment by Industry, 1982

Industry	Total Industry Employment (thousands)	Clerical Employment (thousands)	Clerical Staffing Ratio (percent)	Percent of Total Clerical Employment	Cumulative Percentage of Total Clerical Employment
State and local government and educational services	13,068	2,512	19.2	13.4	13.4
Miscellaneous retail trade	10,476	2,496	23.8	13.3	26.8
Wholesale trade	5,294	1,531	28.9	8.2	34.9
Banking	1,650	1,180	71.5	6.3	41.2
Federal government	2,739	1,138	41.5	6.1	47.3
Insurance	1,700	911	53.6	4.9	52.2
Miscellaneous business services	3,139	896	28.5	4.8	57.0
Hospitals	4,166	666	16.0	3.6	60.5
Social services, museums, and membership organizations	2,755	587	21.3	3.1	63.7
Credit agencies, security and commodity brokers	1,015	577	56.9	3.1	66.8
Legal and miscellaneous services	1,628	560	34.4	3.0	69.7
Telephone and other communication	1,174	529	45.1	2.8	72.6
Physician and dental offices	1,309	394	30.1	2.1	74.7
Construction	3,913	324	8.3	1.7	76.4
Eating and drinking places	4,781	224	4.7	1.2	77.6
Electric services and gas distribution	792	207	26.2	1.1	78.7
Trucking and warehousing	1,206	199	16.5	1.1	79.8
Miscellaneous printing and publishing	846	192	22.8	1.0	80.8
Real estate	986	188	19.1	1.0	81.8
Miscellaneous personal services	1,219	186	15.3	1.0	82.8

SOURCE: Hunt and Hunt (1985a:Table 4.5); based on BLS data.

EMPLOYMENT LEVELS AND OCCUPATIONAL SHIFTS

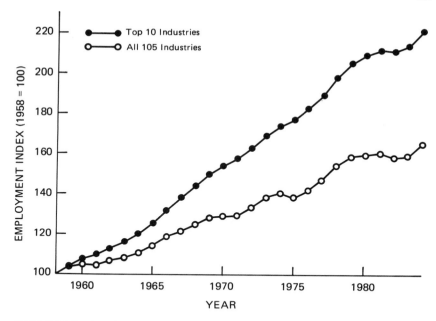

FIGURE 3-3 Total employment in the top 10 clerical-employing industries, 1958-1984. SOURCE: Hunt and Hunt (1985a:Figure 4-A); based on 1958-1984 BLS data.

largely from the fact that employment in these industries does not ordinarily retreat during recessionary periods but continues to expand.

When the employment trends in each of the top 10 clerical-employing industries are considered separately, the diversity among them emerges. Table 3-14 presents the 10 industries ranked by their growth in total employment over the 27-year period. In the fastest-growing industry, miscellaneous business services, the growth rate in employment is particularly striking, more than double the growth rate of the second-ranked industry. Its employment increased almost six times between 1958 and 1982, compared with about 67 percent for all industries. This industry provides a wide variety of services to business firms, including accounting, customized computer software, consulting advice, and temporary personnel placement. Several other industries—banking, credit agencies, security and commodity brokers, and, to a lesser extent, insurance—contributed significantly to clerical job growth during these years. Each of these industries has a staffing ratio for clerical workers in excess of 50 percent, the highest of all industries (see Table 3-13). Hospitals were another important source of clerical employment growth (although the staffing ratio is low, it is a large sector), with rapid growth in total employment over the entire period;

TABLE 3-14 Growth in Employment in the Top 10 Clerical-Employing Industries, 1958–1984

Industry	Total Employment (thousands)						Percentage Growth, 1958–1984
	1958	1980	1981	1982	1983	1984	
Miscellaneous business services	617	3,112	3,334	3,400	3,577	4,033	554
Credit agencies and commodity brokers	340	912	964	991	1,077	1,166	243
Hospitals	908	2,750	2,904	3,014	3,034	2,988	229
Banking	617	1,571	1,628	1,656	1,663	1,684	173
State and local government	5,647	13,375	13,259	13,098	13,099	13,185	133
Social services and museums	1,113	2,099	2,177	2,112	2,117	2,179	96
Wholesale trade	2,980	5,275	5,358	5,278	5,259	5,526	85
Miscellaneous retail trade	6,259	10,452	10,488	10,397	10,592	11,106	77
Insurance	1,021	1,676	1,702	1,714	1,721	1,757	72
Federal government	2,191	2,866	2,772	2,739	2,752	2,783	27
Top 10 industries	23,222	48,714	49,275	49,230	49,898	51,619	122
All 105 industries	60,346	96,653	97,392	95,855	96,540	100,505	67

SOURCE: Calculated from unpublished data provided by H. Allan Hunt and Timothy L. Hunt, Upjohn Institute.

unlike most of the other top 10 industries, however, hospitals experienced actual employment declines between 1983 and 1984.

Employment in the federal government grew very slowly over the entire period, and it has not grown at all since 1967. Alone among the top 10 clerical employers, it experienced slower growth than the national average. Employment growth in state and local government and education, the largest single employer of clerical workers among all industries, was generally above average, but employment has actually declined since 1980.

The foregoing analysis of total employment growth in the top 10 clerical-employing industries and Figure 3-3 suggest that clerical employment has generally increased faster than average employment because it has been concentrated in industries that have grown more than the average. Growth in clerical employment can be more fully analyzed by decomposing it into each of the three factors that contribute to it: (1) overall economic growth, (2) differences in the rates of growth of industries (as noted), and (3) changes in the staffing ratios within industries. Changes in the latter two factors cause an occupation to increase or decrease in *relative* importance in the occupational structure. Changing staffing ratios, which are probably the most visible manifestation of the specific effects of technological change on occupational employment, have also affected clerical employment growth. As will be shown below, some industries have increased their use of clerical workers per dollar of output, while others have reduced their use. The clerical staffing ratios for computer-related occupations, for example, have risen over time in many industries due to the dramatic increases in the use of computers, while the staffing ratios for stenographers have been falling for some time.[4]

An analysis of the clerical employment change for the 1972-1982 period, performed by Hunt and Hunt (1985a), shows that growth of clerical employment at 28.8 percent was somewhat above the economywide average. Of these 28.8 percentage points, the contribution of overall growth was 21.1; that of differential rates of growth among industries, 4.4; and that of changes in occupational staffing ratios within industries, 3.3. That is, clerical employment in the aggregate was boosted by 4.4 percent because industries that were relatively large employers of clerical workers expanded more than other industries. This 4.4 percent, however, represents the average of widely differing contributions that the various industries made to clerical employment growth. As can be seen in Table 3-15, those contributions range from -22.9 to +49.1. An increase in the clerical staffing ratio from 17.4 percent in 1972 to 18.5 percent in 1982 caused a further 3.3 percent rise in clerical employment. This, too, represents the average effect of changes in staffing ratios in all industries. Hence, in the 10 years from 1972 to 1982, neither changing staffing ratios nor differential rates of industry growth were major contributors to overall growth in clerical employment, although both factors were modestly positive during the period; rather, overall economic growth accounted for the bulk of clerical employment growth.

The differential experiences in the various industries shed light on sources of change; an understanding of them is likely to be helpful in considering future employment growth. Table 3-15 shows the decomposition of the growth in clerical employment for each of the 11 major industrial sectors (from Hunt and Hunt, 1985a). In retail trade, for example, there was a total increase of 741,000 clerical workers (or 35.3 percent) in the industry between 1972 and 1982. Economic growth at the same rate as the economy as a whole would have increased clerical employment by 444,000 (21.1 percent) in this industry for this period; the more intensive use of clerical workers in the retail trade sector over the decade increased clerical jobs by 252,000 (12.0 percent); and the retail sector's above-average growth increased clerical employment by 45,000 (2.1 percent).

A particularly striking result shown in this table is that staffing ratios for

---

[4]Staffing ratios may change for reasons other than the use of innovations; organizational change or job title change with no change in job content may occur, for example. It should be understood that any time an individual occupational staffing ratio changes, all of the remaining staffing ratios in that industry will also change. A very large increase in the need for professional or assembly-line workers in an industry could reduce the staffing ratio for clerical workers even though no change affected the nature of their jobs directly.

102

TABLE 3-15 Clerical Employment Growth by Industry, 1972–1982

Industry	Clerical Employment Changes				Decomposition of Clerical Employment Changes, 1972–1982					
					Absolute Changes Due to:			Percent of Employment Changes Due to:		
	1972 Employment (thousands)	1982 Employment (thousands)	Change in Employment 1972–1982 (thousands)	Change in Employment 1972–1982 (percent)	Aggregate Economic Growth (thousands)	Differential Rates of Industry Growth (thousands)	Staffing Ratio Changes (thousands)	Aggregate Economic Growth	Differential Rates of Industry Growth	Staffing Ratio Changes
Agriculture	48	83	35	72.9	10	−11	36	21.1	−22.9	75.0
Mining	59	128	69	116.9	12	29	28	21.1	49.1	47.5
Construction	362	451	89	24.6	77	−44	56	21.1	−12.1	15.5
Durables	1,352	1,513	161	11.9	286	−244	119	21.1	−18.0	8.8
Nondurables	1,040	1,074	34	3.3	220	−222	36	21.1	−21.3	3.5
Utilities	1,307	1,463	156	11.9	276	−23	−97	21.1	−1.8	−7.4
Wholesale trade	684	844	160	23.4	145	86	−71	21.1	12.6	−10.4
Retail trade	2,099	2,840	741	35.3	444	45	252	21.1	2.1	12.0
Finance	2,007	2,750	743	37.0	424	457	−138	21.1	22.8	−6.9
Services	3,691	5,473	1,782	48.3	781	605	396	21.1	16.4	10.7
Public administration	1,678	1,827	149	8.9	355	−53	−153	21.1	−3.2	−9.1
Total	14,326	18,446	4,120	28.8	3,029	625	466	21.1	4.4	3.2

NOTES: Totals and percentages may not add exactly due to rounding. Percentages based on 1972 employment.

SOURCE: Hunt and Hunt (1985a:Table 4.8); based on CPS data.

clerical jobs are falling in a number of industries that employ large numbers of clerical workers. For example, in finance, which has experienced above-average growth in clerical employment over the 10-year period (37.0 in finance compared with 28.8 percent for the whole economy), a decline in the staffing ratio would have reduced clerical employment by 6.9 percent. This decline suggests that the adoption of the new technologies in finance, which has been regarded as a leader, may be having a negative effect on the relative employment of clerical workers. At the same time, the rapid technological change in this industry probably contributed to its rapid growth (through lower prices and expanded services and products), suggesting net positive effects of technology on employment.

Staffing ratios for clerical jobs also fell in three other important industries—utilities, wholesale trade, and public administration. Technology may have been a factor in these declines. In utilities, for example, the employment of telephone operators declined substantially as automatic switching equipment became more advanced. Less is known about developments in wholesale trade. As Hunt and Hunt (1985a) note, the decline in public administration is puzzling. In recent years government has been adopting office automation on a large scale, although it was a slow starter. The postal service has automated numerous clerical jobs in the mail-sorting operation over the past several decades. It is also true that government grew relatively slowly during this time period. Perhaps government administrators, when faced with tight budgets and rising demands for services, economize more on clerical jobs than on other positions.

Interestingly, changes in staffing ratios contributed to increases in clerical employment in those industries that are relatively small employers of clerical workers: agriculture, mining, construction, and durable and nondurable manufacturing. These changes may point to sources of clerical job growth. Clearly, more study of these trends is called for in order to better understand the reasons for changes in staffing ratios.

OUTLOOK FOR CLERICAL EMPLOYMENT

Overall Growth

The critical question for considering future growth in clerical employment involves changes in its three underlying components. How much will the economy grow overall? Will the fast-growing industries that have employed the majority of clerical workers continue to experience above-average growth? How will staffing ratios for clerical occupations in the various industries change? Obviously, projecting the future of these factors is not easy. While definitive conclusions are not possible, however, alternative estimates can reduce uncertainty about possible outcomes.

TABLE 3-16 Bureau of Labor Statistics Projected Occupational Employment Growth, 1982-1995

Occupation	Employment Changes				Decomposition of Employment Changes, 1982-1995						
	1982 Employment (thousands)	1995 Employment (thousands)	Change in Employment 1982-1995 (thousands)	Change in Employment 1982-1995 (percent)	Absolute Changes Due to:				Percent of Changes Due to:		
					Aggregate Economic Growth (thousands)	Differential Rates of Industry Growth (thousands)	Staffing Ratio Changes (thousands)		Aggregate Economic Growth	Differential Rates of Industry Growth	Staffing Ratio Changes	
Professional, technical	15,071	20,177	5,106	33.9	4,228	-99	977		28.1	-0.7	6.5	
Managers, officials	7,696	10,659	2,963	38.5	2,159	162	642		28.1	2.1	8.3	
Sales workers	5,906	7,704	1,798	30.4	1,657	141	0		28.1	2.4	0.0	
Clerical workers	18,717	23,673	4,957	26.5	5,251	295	-588		28.1	1.6	-3.1	
Craft and related workers	10,133	13,223	3,089	30.5	2,843	36	211		28.1	0.4	2.1	
Operatives	12,504	14,896	2,392	19.1	3,508	-566	-550		28.1	-4.5	-4.4	
Laborers, nonfarm	5,572	6,794	1,222	21.9	1,563	-203	-139		28.1	-3.6	-2.5	
Service workers	15,318	19,727	4,408	28.8	4,297	580	-469		28.1	3.8	-3.1	
Total	91,950	117,745	25,795	28.1								

NOTES: Some occupational detail is omitted. Totals and percentages may not add exactly due to omission of some occupational detail and rounding error. Percentages are based on 1972 employment. The 378 OES industries were first aggregated to 105 industries before accomplishing the decomposition. The OES data tape includes wage and salary employment only.

SOURCE: Hunt and Hunt (1985a:Table 5.2); based on data tape from 1982-1995 OES/BLS occupational employment projections.

EMPLOYMENT LEVELS AND OCCUPATIONAL SHIFTS   105

Projections of occupational employment begin with the concept that the demand for labor is a *derived demand*. That is, employers wish to hire workers in order to produce the goods or supply the services that they sell to consumers. As a first step in developing projections, the BLS forecasts aggregate economic activity and industry demand for a given year in the future. In a second step, it estimates total employment by industry, based on forecasts of productivity growth over the period. In a third step, it estimates how many workers in each occupational category will be needed, adjusting the occupational staffing ratios in each industry on a judgmental basis to account for any changes in occupational demand anticipated as a result of technological change or other factors. As Hunt and Hunt (1985a:5.6–7) note:

> Technological change actually enters the system in at least three places. First, the industry output projections should account for anticipated changes in demand induced by technological change. Secondly, the estimated productivity gains forecast for each industry should be influenced by technological change. Finally, the staffing patterns themselves are altered directly to account for technological change. In other words, technological change will have specific effects on some occupations, it will have an overall impact on the productivity of workers, and it will affect the demand for goods and services generally. . . . [T]his system involves a considerable amount of judgment, especially in anticipating the effects of technological change. There are no simple equations that predict changes in staffing ratios within an industry. In fact, the BLS staff has found that trends in industry employment levels can be predicted more accurately than the changes in occupational employment. . . . This is due in large part to the difficulty of projecting specific occupational impacts of technological change.

Table 3-16 summarizes the Hunt and Hunt (1985a:Chapter 5) analysis of the Bureau of Labor Statistics projections for 1995. The projections that Hunt and Hunt worked with used 1982 as the base year and were released by BLS in 1983. More recent projections for 1995, with 1984 as the base year, were released by BLS late in 1985; the panel's necessarily brief analysis of the newer projections appears below.

Hunt and Hunt's analysis of the 1983 BLS projections shows expected employment increases in eight broad occupational categories, including decomposition of their sources of growth.[5] In the past the industries that employed the most clerical workers have been faster growing than the average. That differential began to erode in the 1970s, and it is not expected to reemerge by 1995. Differential rates of industry growth are expected to increase clerical employ-

---

[5]The BLS does not actually forecast occupational employment growth at the major group level, but it is still helpful to analyze the projections at this level of aggregation to provide an overview of the system. It also enables us to compare those projections with the historical CPS data reviewed above.

ment by only 1.6 percent between 1982 and 1995 (see Table 3-16), compared with 4.4 percent between 1972 and 1982.

Of the large clerical-employing industries, growth is expected to be smallest in state and local government and education and in the federal government. Employment by the federal government is not expected to increase at all, while employment in state and local government and education is expected to recover from the declines that occurred in the 1980 and 1982 recessions and to grow, but significantly more slowly than the national average. The fastest-growing industries among the top employers of clerical workers are expected to be credit agencies and commodity brokers, hospitals, miscellaneous business services, and banking (Hunt and Hunt, 1985a). Of these, the growth anticipated by BLS for banking and hospitals is surprising. As noted in Chapter 2, both banking and health services are experiencing significant structural change. In banking, factors that have contributed to speculation that employment growth may slow are deregulation, the increased use of automatic teller machines, the closing of branch and satellite banks, and declines in employment among several of the largest banks in the nation. In health services, employment is shifting away from hospitals toward other providers, and employment growth overall seems to be slowing.

In these projections, average employment growth is expected to be 28.1 percent; only three of the eight broad occupations shown in Table 3-16 are expected to grow slower than the average—operatives, laborers, and clerical workers, with clerical workers showing only slightly less than average growth. The variation of growth rates for the occupations around the average growth rate is relatively small, from 19.1 to 38.5 percent, although the range actually experienced between 1972 and 1982 was from -5.9 to $+46.9$ percent. And between 1972 and 1979 (omitting the recession years, which might distort the data), growth rates ranged from 8.5 to 35.4 percent. The Hunt and Hunt (1985a) analysis suggests that BLS projects less change in the relative importance of broad occupations over the 13 years from 1982 to 1995 than actually occurred during the 7 years from 1972 to 1979. This finding probably also reflects BLS estimation methods, which make it difficult to project dramatic shifts, a natural caution that is probably not misplaced. In keeping with these results, the decomposition shown in Table 3-16 indicates that the relative effects of changing staffing ratios and differential rates of industry growth are small for all occupations. The decomposition does reveal a significant change, however: the impact of staffing ratios on clerical employment is expected to be negative, the only reversal projected by BLS from the existing trends in the historical data.

In its later projections for 1995 (released in November 1985), BLS reduced its estimates of overall growth and further reduced the expected relative size of clerical employment (Silvestri and Lukasiewicz, 1985). In the 1985 projections of 1995 employment (which use 1984 rather than 1982 as the base year), employment growth over the 11-year period is anticipated at 15 percent in the moderate trend series (meaning economic growth would be moderate) in comparison with 25 percent in the earlier projection for the 13-year period, and the growth of clerical workers is expected to be 9.5 percent, well below the economywide average. The share of clerical workers in total employment would thus fall from 17.5 percent of the labor force in 1984 to 16.7 percent in 1995. In their discussion of the new projections, Silvestri and Lukasiewicz point out that growth in clerical employment was about average between 1973 and 1984 and that, despite slower projected growth, clerical employment would still remain the largest category of employment and would add 1.8 million jobs. They suggest that the main reason for the reduced projection is revision in the bureau's estimate of the impact of office automation, not only its direct impact on the tasks performed by clerical workers but also on a shift in tasks from clerical workers to professional, technical, and managerial workers (which are expected to be especially fast-growing occupational categories).

In the earlier (1983) projections, the growth of employment in clerical work, as analyzed by Hunt and Hunt (see Table 3-16), was expected to be only 1.6 percent less than overall employment growth (26.5 compared with 28.1 for the 1982–1995 period). In the later (1985) projections, as analyzed by Silvestri and Lukasiewicz, the difference is much more dramatic: clerical employment is expected to grow 9.5 percent from 1985 to 1995, compared with a growth in total employment of 14.9 percent; BLS now projects clerical work to grow 5.4 percent less than—and only about two-thirds as much as—total employment. Although the lack of time prevented a thorough analysis of these new projections similar to that performed by Hunt and Hunt on the earlier ones, a first-cut analysis suggests that the dramatic projection of much slower than average growth in clerical jobs is actually largely the result of classification changes: for example, cashiers have been reclassified as sales workers. And because BLS projects the occupation of cashiers to grow very rapidly (more rapidly than projected in 1983), removing them—a very large category—from the clerical classification has the effect of slowing the growth rate of clerical workers more than would have occurred without the change.

A comparison of the 1983 and 1985 BLS projections appears in Table 3-17 along with some alternative projections. The 1983 projections are taken from Hunt and Hunt (1985a) and are based on the OES tape; they are not the same as the published data, since the tape excludes unpaid family workers and self-employed people. The 1985 BLS projections have been adjusted to correspond to the Hunt and Hunt 1983 version; the data are from the tape, and the panel has restored the classification (approximately) to its 1983 version. The category of clerical workers was made more similar to its 1983 form: cashiers, 1,570,000 in 1982 and projected to be 2,469,000 in 1995 (Silvestri and Lukasiewicz,

TABLE 3-17 Projections of Occupational Employment Growth, 1982–1995

Projection	BLS[a] 1983	BLS[b] 1985	Leontief[c] -Duchin	Panel[d] 1983	Panel[d] 1985
Clerical employment					
1982	18,717	18,717	18,032	18,717	18,717
1995	23,673	22,680	17,786	21,641	20,674
Difference (1982–1995)	4,957	3,963	−246	3,167	1,957
Percent change (1982–1995)	26.5	21.2	−1.4	16.9	10.5
Total employment					
1982	91,950	91,950	107,284	91,950	91,950
1995	117,745	112,361	143,753	117,745	112,341
Percent change (1982–1995)	28.1	22.2	34.0	28.1	22.2
Clerical workers as a percent of total					
1982	20.4	20.4	16.8	20.4	20.4
1995	20.1	20.2	12.4	18.4	18.4
Percent change (1982–1995) (clerical staffing ratio)	−1.5	−1.0	−26.2	−9.9	−9.9

[a]1983 projections from BLS, as presented in Hunt and Hunt (1985a).
[b]1985 projections from BLS, unpublished data, adjusted by panel to correspond to 1983 BLS occupational classifications.
[c]Data from Leontief and Duchin (1984), as presented in Hunt and Hunt (1985a).
[d]Computed by the panel using a larger decrease in the clerical proportion of total employment, but retaining the other projections from BLS for the respective years.

NOTE: Data for both 1983 and 1985 Bureau of Labor Statistics projections are from the unpublished Occupational Employment Statistics tape and differ from published figures because unpaid family workers and the self-employed are not included here.

1985), were added back, along with shipping packers, another fairly large category that had been removed, while order fillers were removed from the 1985 projections since they had been added to the clerical category between 1983 and 1985.

As Table 3-17 shows, the result of this adjustment is that, overall, BLS has not altered its estimate of clerical employment growth. As in 1983, clerical employment growth is expected to be slightly less than average, so that the proportion of clerical employment to total employment falls slightly (indicating a small negative staffing ratio, shown on the last row in the table). However, between 1983 and 1985, BLS changed some of its employment projections for specific clerical jobs substantially. Apparently, forecast increases (e.g., for cashiers) almost offset forecast decreases (e.g., for secretaries). These changes in specific occupations are examined further below.

The BLS projections on balance suggest substantial growth in clerical work. Other forecasts have projected a much more pessimistic outlook for clerical employment. Perhaps the best known of these is the 1984 study by Leontief and Duchin, which can be seen in comparison with the BLS 1983 and 1985 projections in Table 3-17. While BLS projects increases in clerical employment over the period only slightly below the national average, Leontief and Duchin predict an actual decrease of 1.4 percent in clerical employment for the 1982–1995 period. Leontief and Duchin also predict relatively slow growth in employment of managers, 11.3 percent, in comparison with the above-average 38.5 percent 1983 BLS projections. If Leontief and Duchin are correct, even in part, it could mean not only displacement for large numbers of clerical workers but also difficulties for managers and other workers seeking to enter management positions in the office.

It is highly likely, however, that the Leontief-Duchin projections overestimate the negative effect of technology on clerical employment. For example, Leontief and Duchin project a 7.2 percent decline in secretarial employment, in comparison with the 30 percent increase forecast by BLS in 1983 (or the 26 percent increase forecast by BLS in 1985). The Leontief-Duchin estimate is based on the study with the largest estimate of productivity gains resulting from use of word-processing equipment, even though other studies suggest considerably smaller effects. They further assume that word-processing equipment creates no "new" work, such as repeated drafts. They also assume fairly rapid diffusion of the new technology and do not take into account that productivity gains may vary by industry, being highest where the work to be done is routine and repetitive and lower where it tends to be more unique and less repetitive. Finally, they consider no technological changes over the period other than computer-based technologies, distorting their estimates of the relative importance of clerical jobs. The overall decline in the staffing ratio for clerical workers projected by Leontief and Duchin, 26.2 percent, is very large, even more than the decrease for operatives of 22.3 percent actually experienced between 1972 and 1982 (calculated from data in Hunt and Hunt, 1985a).

Are Leontief and Duchin right about their projected large decline in clerical employment? The panel thinks not. Comparison between the largest actual decline in a staffing ratio, that for operatives, and the Leontief and Duchin projection for clerical workers can illuminate the issue. First, the decline observed for operatives has been occurring for some time, while Leontief and Duchin predict a rapid about-face for clerical work. Second, the trend observed for operatives in the 1972–1982 period includes the effect of the 1981–1982 recession, the steepest since the Great Depression of the 1930s. Since blue-collar employment falls steeply during a recession, the large change observed for operatives probably reflects this cyclical effect. Such a cyclical effect may not recur, and, even if it did, it would likely be smaller for clerical workers, whose employment traditionally does not fall as much as that of blue-collar workers in a recession. (Indeed, only in the last recession, 1981–1982, did clerical employment actually decline; in all others it continued to grow.) Third, the drop in operatives may also be due to a classification change, which resulted in some operatives being reclassified as technicians. Hence, the actual drop in the employment of workers who carry out those functions may not have been as great as the numbers indicate.[6] For all these reasons, the large decline that Leontief and Duchin predict in the staffing ratio of clerical workers seems highly unrealistic. It is worth noting, however, that, if the secretarial function actually disappears, or declines dramatically, as Leontief and Duchin predict, then employment opportunities for women will become a serious problem, especially if their training is not easily transferable to other jobs.

As an illustration of the effects of staffing ratio changes that are large but more plausible than Leontief and Duchin's, the panel has computed two alternative projections, corresponding to the 1983 and 1985 BLS projections. These projections use a larger decrease in the clerical staffing ratio than BLS forecasts, but retain BLS figures for projected total employment. As noted above, several industries experienced declines in the clerical staffing ratio between 1972 and 1982; the largest rate of decline occurred in public administration and wholesale trade. The 9.9 percent decline in the clerical staffing ratio shown in the bottom row of Table 3-17 for the 1982–1995 period is computed by applying the annual rate of decline in the clerical staffing ratio actually experienced by these two industries between 1972 and 1982 to the whole economy. As can be seen in Table 3-17, under this assumption, clerical employment increases by 16.9 percent in the 1983-based projections and by 10.5 percent in the 1985-based projections. These increases are sizable, even though they are considerably below the average for the economy as a whole. (The primary cause of the difference in the two projections is the slower overall employment growth estimated by BLS in 1985 than in 1983.) Moreover, it might be argued that even this figure could be an overestimate of the reduction in growth in these jobs, since the rate of decline in the staffing ratios in public administration and wholesale trade may very well have been influenced by factors other than technological change. Of course, it could be an underestimate if the rate of technological change were to accelerate over the next 10 years. Although the growth in clerical employment in the panel's estimates is much larger than that projected by Leontief and Duchin, it is also substantially smaller than that projected by BLS because the change in the clerical staffing ratio is substantial and negative. In the panel's judgment, these estimates reflect the "most plausible

---

[6]Similar shifts are occurring in the classification of clerical workers, with uncertain effects on observable future employment. One possibility is that Leontief and Duchin's predictions will be supported by future observations of decline in such clerical categories as secretaries, but that the decline may turn out to be an artifact of reclassification. If all that happens is that secretaries take on other titles, a decline in the category "secretary" will not significantly affect employment opportunities for women, as they will be able to find employment in newly titled but fundamentally the same jobs.

worst case"—as large a negative effect for clerical employment as is likely to occur.

Based on this analysis and its best judgment concerning various trends, the panel concludes that the rate of growth of clerical employment will decline, most probably falling somewhere between the 1985 BLS and the 1985 panel estimates. In the panel's judgment, there is insufficient evidence to support the much more negative outlook of the Leontief-Duchin model. The panel does not foresee dramatic declines in the need for most clerical functions. Indeed, those that emphasize human interaction are likely to continue to grow. The panel does not, therefore, foresee a major unemployment problem for clerical workers. It is interesting to note that in its 1985 projections for 1995, reflecting its latest thinking, BLS approaches the panel's previous most plausible worst case forecast; BLS's revised 1985 projections for clerical employment fall between the 1983 BLS and the 1983 panel projections. As Table 3-17 shows, the 1985 BLS projection is 22.7 million clerical jobs in 1995 compared with the panel's projection of 20.7 million. In all the panel and BLS estimates, at least 2 million clerical jobs will be created by 1995, while Leontief and Duchin project a loss of 250,000 clerical jobs.

Although the number of clerical jobs will continue to grow overall, slower aggregate growth, and the inevitable shifts that will occur among the many subfields of clerical work, may well cause some structural unemployment. Projections of some of the more detailed clerical occupations are examined next to see what occupational shifts can be expected.

OCCUPATIONAL SHIFTS

The 1983 BLS projections for 95 specific clerical occupations, as analyzed by Hunt and Hunt (1985a), are shown in Table 3-18, ranked in order of clerical staffing ratio change. Projected employment growth varies from +76.1 percent for computer operators over the 13-year period (1982 to 1995) to −20.0 percent for central office telephone operators. The proportion of the employment change that can be attributed to changes in staffing ratios, the more or less intensive use of specific occupations, varies from +38.4 for computer operators to −55.6 for central office telephone operators. Clearly, the diversity of change anticipated in clerical suboccupations is large.

In the OES classification, the three largest occupations are general office clerks, secretaries, and cashiers. No negative effects of technological change on employment trends in these occupations are discernible in these projections. The staffing ratio for cashiers is expected to increase significantly—despite the increased use of "intelligent cash registers"—contributing to the overall 48.2 percent growth forecast for that occupation. The other two occupations are expected to experience only slightly more growth than the average growth for

TABLE 3-18 Bureau of Labor Statistics Projected Occupational Employment Growth, 1982–1995, Detailed Clerical Occupations, Ranked by Percent of Employment Change Due to Staffing Ratio Changes

Occupation	Employment Changes				Decomposition of Employment Changes, 1982–1995				Percent of Changes Due to:		
	1982 Employment (thousands)	1995 Employment (thousands)	Change in Employment 1982–1995 (thousands)	Change in Employment 1982–1995 (percent)	Absolute Changes Due to:						
					Aggregate Economic Growth (thousands)	Differential Rates of Industry Growth (thousands)	Staffing Ratio Changes (thousands)		Aggregate Economic Growth	Differential Rates of Industry Growth	Staffing Ratio Changes
All clerical workers	**18,716.6**	**23,673.5**	**4,956.9**	**26.5**	**5,250.6**	**294.8**	**−588.4**		**28.1**	**1.6**	**−3.1**
Computer operators	210.0	369.7	159.7	76.1	58.9	20.1	80.7		28.1	9.6	38.4
Claims adjusters	65.4	97.6	32.1	49.1	18.4	−4.0	17.8		28.1	−6.2	27.3
Insurance checkers	14.9	22.4	7.4	49.8	4.2	−0.3	3.5		28.1	−2.0	23.7
Peripheral EDP equipment operators	47.7	78.6	30.8	64.6	13.4	6.6	10.9		28.1	13.7	22.8
Telephone ad takers, newspapers	10.4	14.5	4.2	40.5	2.9	−0.9	2.2		28.1	−8.8	21.2
Claims clerks	63.0	89.8	26.8	42.5	17.7	−4.2	13.3		28.1	−6.7	21.1
Credit authorizers	20.2	30.5	10.3	51.2	5.7	0.6	4.0		28.1	3.1	20.0
Worksheet clerks	10.6	15.3	4.7	44.1	3.0	−0.2	1.9		28.1	−2.0	18.1
Service clerks	23.6	34.9	11.3	48.1	6.6	0.8	4.0		28.1	3.2	16.9
Cashiers	1,532.4	2,270.5	738.1	48.2	429.9	56.6	251.6		28.1	3.7	16.4
Insurance clerks, medical	85.7	139.1	53.4	62.2	24.1	15.9	13.4		28.1	18.5	15.7
Teachers' aides	462.7	593.1	130.3	28.2	129.8	−69.4	70.0		28.1	−15.0	15.1
Credit clerks, banking and insurance	49.6	76.4	26.8	54.0	13.9	5.5	7.4		28.1	11.1	14.9

Adjustment clerks	33.8	47.4	13.6	40.1	9.5	0.3	3.8	28.1	0.9	11.2
Telegraph operators	4.4	6.4	2.0	46.1	1.2	0.3	0.5	28.1	7.5	10.6
Transportation agents	20.6	28.1	7.5	36.3	5.8	−0.1	1.8	28.1	−0.6	8.9
Mortgage closing clerks	15.3	22.6	7.2	47.2	4.3	1.7	1.3	28.1	10.8	8.4
Real estate clerks	16.6	23.5	6.9	41.8	4.7	1.0	1.2	28.1	6.2	7.5
Customer service representatives	88.9	123.8	34.8	39.2	25.0	3.4	6.5	28.1	3.8	7.3
Receptionists	381.1	569.7	188.6	49.5	106.9	54.2	27.5	28.1	14.2	7.2
Claims examiners, insurance	47.3	62.1	14.9	31.5	13.3	−0.9	2.6	28.1	−2.0	5.4
Raters	52.6	69.0	16.4	31.1	14.8	−1.1	2.7	28.1	−2.0	5.0
Admissions evaluators	10.5	12.1	1.6	15.4	2.9	−1.8	0.5	28.1	−17.3	4.7
Loan closers	45.3	64.0	18.8	41.5	12.7	4.2	1.9	28.1	9.2	4.2
Circulation clerks	9.5	11.8	2.3	23.8	2.7	−0.8	0.4	28.1	−8.4	4.2
Insurance clerks, except medical	10.6	14.6	4.0	37.6	3.0	0.6	0.4	28.1	5.7	3.9
Clerical supervisors	466.1	627.4	161.3	34.6	130.7	13.4	17.2	28.1	2.9	3.7
Library assistants	80.2	94.6	14.4	18.0	22.5	−10.8	2.7	28.1	−13.4	3.4
Personnel clerks	102.3	131.0	28.7	28.0	28.7	−3.3	3.3	28.1	−3.3	3.2
Title searchers	5.1	7.1	2.0	38.5	1.4	0.4	0.2	28.1	7.4	3.1
Payroll and timekeeping clerks	201.2	268.8	67.6	33.6	56.4	6.6	4.5	28.1	3.3	2.2
Mail clerks	98.7	129.7	31.0	31.4	27.7	1.2	2.2	28.1	1.2	2.2
Procurement clerks	46.9	59.0	12.2	25.9	13.2	−1.9	0.9	28.1	−4.1	2.0
Rate clerks, freight	10.2	12.5	2.3	22.6	2.9	−0.7	0.2	28.1	−7.2	1.8
Statement clerks	33.6	44.2	10.7	31.7	9.4	0.8	0.4	28.1	2.3	1.3
Proofreaders	16.2	20.6	4.3	26.8	4.5	−0.4	0.2	28.1	−2.6	1.3
Production clerks	199.8	260.0	60.2	30.1	56.0	1.9	2.2	28.1	1.0	1.1

Continued

TABLE 3-18 (Continued)

	Employment Changes				Decomposition of Employment Changes, 1982–1995					
					Absolute Changes Due to:			Percent of Changes Due to:		
Occupation	1982 Employment (thousands)	1995 Employment (thousands)	Change in Employment 1982–1995 (thousands)	Change in Employment 1982–1995 (percent)	Aggregate Economic Growth (thousands)	Differential Rates of Industry Growth (thousands)	Staffing Ratio Changes (thousands)	Aggregate Economic Growth	Differential Rates of Industry Growth	Staffing Ratio Changes
Switchboard operators/ receptionists	203.8	281.6	77.9	38.2	57.2	18.5	2.2	28.1	9.1	1.1
Town clerks	26.0	29.1	3.1	11.7	7.3	−4.5	0.3	28.1	−17.3	1.0
Meter readers, utilities	30.5	37.9	7.3	24.0	8.6	−1.5	0.3	28.1	−4.9	0.9
General clerks, office	2,342.0	3,037.4	695.5	29.7	657.0	20.6	17.8	28.1	0.9	0.8
Traffic agents	17.8	22.3	4.5	25.1	5.0	−0.6	0.1	28.1	−3.3	0.4
Weighers	24.3	28.7	4.3	17.8	6.8	−2.6	0.1	28.1	−10.5	0.3
Safe deposit clerks	13.9	18.1	4.2	30.5	3.9	0.3	0.0	28.1	2.4	0.0
Dispatchers, police, fire, and ambulance	47.8	53.4	5.5	11.6	13.4	−7.9	0.0	28.1	−16.5	0.0
Order clerks	257.0	325.4	68.4	26.6	72.1	−3.0	−0.7	28.1	−1.2	−0.3
Customer service representatives, printing and publishing	8.4	10.3	1.9	22.2	2.4	−0.5	0.0	28.1	−5.4	−0.5
All other clerical workers	1,220.5	1,542.0	321.6	26.3	342.4	−14.0	−6.8	28.1	−1.1	−0.6

115

Occupation								
Bookkeeping, billing machine operators	171.5	221.7	50.2	29.3	48.1	3.4	−1.3	28.1	2.0	−0.8
All other office machine operators	89.0	121.8	32.8	36.8	25.0	8.8	−1.0	28.1	9.9	−1.1
Dispatchers, vehicle services or workers	86.9	109.7	22.8	26.3	24.4	0.1	−1.7	28.1	0.2	−2.0
Collectors, bill and account	90.9	130.9	40.0	44.0	25.5	16.5	−2.0	28.1	18.1	−2.2
Tellers	471.5	613.1	141.6	30.0	132.3	20.1	−10.7	28.1	4.3	−2.3
Secretaries	2,298.7	2,988.5	689.8	30.0	644.8	98.3	−53.3	28.1	4.3	−2.3
Court clerks	27.3	29.4	2.2	7.9	7.7	−4.7	−0.8	28.1	−17.3	−2.9
Proof machine operators	47.4	59.4	11.9	25.2	13.3	0.3	−1.6	28.1	0.6	−3.4
Sorting clerks, banking	7.4	9.3	1.9	25.5	2.1	0.1	−0.3	28.1	1.5	−4.1
Checking clerks	18.0	22.7	4.7	26.2	5.0	0.5	−0.8	28.1	2.7	−4.5
Shipping packers	339.0	402.1	63.1	18.6	95.1	−15.2	−16.8	28.1	−4.5	−5.0
Messengers	49.7	65.4	15.8	31.8	13.9	4.6	−2.7	28.1	9.2	−5.5
Transit clerks	7.3	8.9	1.6	22.6	2.0	0.1	−0.5	28.1	0.9	−6.4
Welfare investigators	11.8	12.3	0.5	4.0	3.3	−2.0	−0.8	28.1	−17.1	−7.0
Shipping and receiving clerks	364.3	430.4	66.1	18.2	102.2	−7.4	−28.7	28.1	−2.0	−7.9
Eligibility workers, welfare	31.5	32.1	0.6	2.0	8.8	−5.4	−2.8	28.1	−17.1	−9.0
Stock clerks, stockroom and warehouse	827.3	983.5	156.3	18.9	232.1	0.9	−76.7	28.1	0.1	−9.3
Car rental clerks	16.2	21.6	5.4	33.3	4.6	2.5	−1.6	28.1	15.1	−9.9

Continued

TABLE 3-18 (Continued)

Occupation	Employment Changes				Decomposition of Employment Changes, 1982–1995						
					Absolute Changes Due to:				Percent of Changes Due to:		
	1982 Employment (thousands)	1995 Employment (thousands)	Change in Employment 1982–1995 (thousands)	Change in Employment 1982–1995 (percent)	Aggregate Economic Growth (thousands)	Differential Rates of Industry Growth (thousands)	Staffing Ratio Changes (thousands)		Aggregate Economic Growth	Differential Rates of Industry Growth	Staffing Ratio Changes
Coin machine operators and currency sorters	5.0	6.0	0.9	18.2	1.4	0.0	−0.5		28.1	0.4	−10.3
Desk clerks, except bowling floor	85.3	104.3	19.0	22.3	23.9	5.0	−9.9		28.1	5.8	−11.6
Accounting clerks	728.7	850.0	121.3	16.7	204.4	6.2	−89.3		28.1	0.9	−12.3
Typists	974.9	1,128.8	153.9	15.8	273.5	2.0	−121.6		28.1	0.2	−12.5
Postal mail carriers	234.1	222.7	−11.4	−4.9	65.7	−46.8	−30.3		28.1	−20.0	−12.9
Travel counselors, auto club	5.4	5.9	0.5	9.1	1.5	−0.3	−0.8		28.1	−5.1	−13.9
Protective signal operators	6.9	11.7	4.8	69.4	1.9	3.8	−1.0		28.1	55.6	−14.3
License clerks	5.7	5.5	0.2	−4.0	1.6	−1.0	−0.8		28.1	−17.3	−14.7
Bookkeepers, hand	884.8	1,042.5	157.7	17.8	248.2	40.4	−130.9		28.1	4.6	−14.8
New-accounts tellers	67.3	79.9	12.6	18.8	18.9	4.1	10.3		28.1	6.0	−15.3
Policy-change clerks	27.6	30.5	2.9	10.5	7.7	−0.6	−4.3		28.1	−2.0	−15.6
Traffic clerks	7.1	10.5	3.3	47.0	2.0	2.5	−1.2		28.1	35.8	−16.9

Occupation								
Ticket agents	49.3	48.9	−0.4	−0.7	13.8	−5.2	−8.9	28.1	−10.6	−18.2
Switchboard operators	169.6	211.3	41.7	24.6	47.6	25.7	−31.6	28.1	15.1	−18.6
Reservation agents	52.9	54.9	2.0	3.7	14.8	−3.0	−9.9	28.1	−5.7	−18.6
Statistical clerks	96.1	110.8	14.7	15.3	27.0	5.7	−18.0	28.1	5.9	−18.7
Desk clerks, bowling floor	15.4	17.8	2.4	15.4	4.3	1.1	−3.0	28.1	7.0	−19.7
Directory assistance operators	37.5	43.1	5.6	14.9	10.5	2.8	−7.7	28.1	7.5	−20.6
Duplicating machine operators	36.1	42.3	6.2	17.1	10.1	3.8	−7.8	28.1	10.6	−21.5
Survey workers	51.4	76.1	24.8	48.2	14.4	21.7	−11.4	28.1	42.3	−22.1
Brokerage clerks	16.5	20.3	3.8	23.0	4.6	3.1	−3.9	28.1	18.5	−23.5
Credit reporters	15.3	20.5	5.2	34.4	4.3	4.7	−3.7	28.1	30.8	−24.5
Postal service clerks	306.5	251.8	−54.8	−17.9	86.0	−61.3	−79.5	28.1	−20.0	−25.9
File clerks	293.0	309.5	26.5	9.1	82.2	21.3	−77.0	28.1	7.3	−26.3
Stenographers	265.6	244.9	−20.7	−7.8	74.5	−7.8	−87.4	28.1	−2.9	−32.9
In-file operators	5.0	6.9	1.9	38.8	1.4	2.8	−2.2	28.1	55.6	−44.9
Data-entry operators	318.7	284.6	−34.1	−10.7	89.4	30.6	−154.1	28.1	9.6	−48.4
Purchase and sales clerks, security	5.2	4.9	−0.3	−5.5	1.5	1.0	−2.7	28.1	18.5	−52.0
Central office operators	108.7	86.9	−21.8	−20.0	30.5	8.1	−60.4	28.1	7.5	−55.6

NOTES: Some occupational detail is omitted. Totals and percentages may not add exactly due to omission of some occupational detail and rounding error. Percentages are based on 1972 employment. The 378 OES industries were first aggregated to 105 industries before accomplishing the decomposition. The OES data tape includes wage and salary employment only.

SOURCE: Hunt and Hunt (1985a: Table 5.5); based on data tape from the 1982–1995 OES/BLS occupational employment projections.

all clerical workers, with a change in the staffing ratio having a slightly negative effect for secretaries (as it does for all clerical workers in this projection) and an even smaller positive effect for general office clerks. In the 1983 BLS estimates, the projected growth rates in these two large clerical occupations, and for clerical workers generally, are approximately equal to those projected for total employment. The projected growth for secretaries is entirely consistent with the historical data (except for the apparent decline of secretaries in 1982, which may have been caused by the recession). In the 1985 projections, BLS reduced the estimated growth of secretaries from 1984 to 1995 to 9.5 percent, identical to the growth projected for all clerical work. The projected growth of secretaries and clerical employment generally is only about two-thirds the growth rate projected for total employment. The new estimate represents a decline of about 100,000 secretaries from the 1983 projection, but it would still result in an increase of almost 300,000 secretaries by 1995. Clearly the economy will still need new workers trained in secretarial skills.

In the 1983 projections, the fastest-growing clerical jobs are computer operators, peripheral electronic data processing (EDP) equipment operators, medical insurance clerks, credit clerks in banking and insurance, credit authorizers, insurance checkers, receptionists, claims adjusters, cashiers, and survey workers. Employment growth in these occupations is expected to range from 48.2 to 76.1 percent between 1982 and 1995. Many of these occupations are expected to experience staffing ratio changes equivalent to employment increases of 20 percent or more. This list reflects in part the obvious technological impacts of computers, but it also reflects the continuing or increased importance of interaction between a worker and a customer being served. Cashiers, for example, may replace other sales workers. As Hunt and Hunt put it (1985a:5–17): "A world of both high-tech and high touch is anticipated." Although various electronic office technologies have the capacity to replace some aspects of human interaction—for example, by automatic bank tellers, partially automated telephone number announcements, or computerized ad takers at newspapers—customers may be resistant to using these devices, or the variety in transactions may make them less widely applicable than now anticipated. As Hunt and Hunt note, the fact that such devices can be developed does not guarantee that they will be, or that they will prove to be profitable if they are developed.

The occupations expected to experience the largest percentage declines by 1995 are central office telephone operators, postal service clerks, data-entry operators, stenographers, security workers and purchase and sales clerks, and postal mail carriers. The effects of staffing ratio changes are expected to be large and negative in these occupations. Projected declines in employment for postal mail carriers (a loss of 11,000 jobs) and postal service clerks (a loss of 55,000 jobs) reflect both the large, negative effects of declining staffing ratios (largely due to technological change) and well below average industry growth (probably due to increased competition from other forms of communication services and other mail services). The effect of a large negative staffing ratio change is expected to reduce employment for data-entry operators (despite a positive industry effect) by about 10 percent, or 34,000 jobs. This decline reflects both the use of new technologies to perform the same work (for example, optical character reading) and the capability of new technologies to shift work to others (for example, consumers or professional and managerial staff). Other occupations expected to experience large, negative effects of staffing ratio changes include in-file operators, file clerks, credit reporters, brokerage clerks, and survey workers. Some of these occupations will nevertheless experience positive employment growth because of strong industry demand and overall economic growth. The number of survey workers, for example, is expected to grow substantially (48.2 percent), although the effect of staffing ratio changes alone would be negative (-22.1 percent). It seems likely that rapid technological change is contributing to increased demand for surveys (along with decreased labor input per dollar of output). The occupations of credit reporters (34.4 percent growth) and brokerage clerks (23.0 percent growth) may reflect similar changes. Other occupations that will experience slow growth, such as file clerks, are simply continuing a decline begun in the 1960s or 1970s. Many of the declining and slow growing occupations are back-office jobs that require little or no direct contact with the customer and may have ready technological explanations: file clerks, stenographers, data-entry operators, and central office telephone operators. Many of these occupations are held disproportionately by minority women—postal service clerks, file clerks, data-entry operators, and telephone operators.

These anticipated shifts among clerical suboccupations make it clear that the slowdown in projected growth for clerical workers will not affect all such workers equally. Some will be in even greater demand, some less so. Some occupations will decrease, but it is important to note that none of the absolute decreases shown in Table 3-18 is expected to be especially large.

A preliminary comparison of 1983 and 1985 BLS projections for 1995 employment in detailed clerical occupations shows that in several occupations BLS has projected larger declines than those suggested by the slowdown in employment growth overall (which is a 3.4 percent decrease, based on data published in Silvestri et al., 1983, and in Silvestri and Lukasiewicz, 1985). In many occupations, however, greater growth is now projected. But comparisons are extremely difficult because of changes in the classification of OES data on which the projections are based. Table 3-19 shows comparisons for those occupations that could be straightforwardly matched and in which the projected changes in 1995 employment were significant and negative. The largest percentage changes between the 1983 and 1985 projections include stenographers, statistical clerks, directory assistance operators, central office operators, and

TABLE 3-19 Clerical Suboccupations with Largest Negative Employment Changes in Bureau of Labor Statistics Projections for 1995 Between 1983 and 1985

Occupation (1)	Actual 1982 (2)	1984 (3)	Projection for 1995 1983 (4)	Projection for 1995 1985 (5)	Difference (5)–(4) (6)	Difference as Percentage of 1983 Projection [(5)–(4)]/(4) (7)
Stenographers	270,000	239,000	250,000	143,000	–107,000	–42.8
Statistical clerks	98,000	93,000	114,000	81,000	–33,000	–28.9
Directory assistance operators	38,000	32,000	43,000	30,000	–13,000	–27.9
Payroll and timekeeping clerks	202,000	207,000	269,000	196,000	–73,000	–27.1
Central office operators	109,000	77,000	87,000	68,000	–19,000	–21.8
Tellers	471,000	493,000	613,000	517,000	–96,000	–15.7
Typists	990,000	991,000	1,145,000	1,002,000	–143,000	–12.5
Switchboard operators	279,000	347,000	498,000	447,000	–51,000	–10.2
File clerks	295,000	289,000	321,000	296,000	–25,000	–7.8
Computer operators	211,000	241,000	371,000	353,000	–18,000	–4.9
Teachers' aides	463,000	479,000	593,000	566,000	–27,000	–4.5
Secretaries	2,441,000	2,797,000	3,161,000	3,064,000	–97,000	–3.1

SOURCE: Calculated from data in Silvestri et al. (1983) and Silvestri and Lukasiewicz (1985), adjusted for consistency by the panel.

payroll and timekeeping clerks. For secretaries, teachers' aides, and computer operators, the lower levels of 1995 employment are in keeping with BLS estimates of slower growth overall.

Which clerical occupations are likely to experience the largest increases in employment, especially for women? The latest published projections (Silvestri and Lukasiewicz, 1985) combined with data on the proportion of females in each occupation from the 1980 census yield the employment increases for women shown in Table 3-20. As the table shows, several of the slower-growing, but large, female clerical occupations provide the largest job growth: secretaries, general office clerks, and bookkeeping, accounting, and auditing clerks. A fair degree of uncertainty surrounds all these projections, however. The impact of technological change could be either more or less than is now supposed. Projections for many occupations have been substantially changed by BLS between 1983 and 1985, generally downward. Substantial absolute declines are projected for stenographers (87,000 jobs for women, a 40.3 percent decline), postal service clerks (10,000 jobs, an 8.5 percent decline), statistical clerks (9,000 jobs, a 12.7 percent decline), payroll and timekeeping clerks (9,000 jobs, a 5 percent decline), and central office operators (8,000 jobs, an

TABLE 3-20 Clerical Occupations with Largest Projected Job Growth for Women, 1984–1995

Occupation (and Percent Female[a])	Employment[b] (thousands) 1984	1995	Change in Female Employment,[c] 1984–1995 Number (thousands)	Percent
Secretaries (98.8)	2,797	3,064	265	9.6
General office clerks (82.1)	2,398	2,629	190	9.6
Bookkeeping, accounting, and auditing clerks (89.7)	1,973	2,091	106	6.0
Switchboard operators (91.0)	347	447	91	28.7
Teachers' aides and educational assistants (92.7)	479	566	82	18.3
Receptionists and information clerks (93.4)	458	542	78	18.2
Computer operators, excluding peripheral equipment operators (58.9)	241	353	65	46.1
Order clerks, material, merchandise and services (67.4)	297	355	38	19.2
Billing, posting, and labeling machine operators (87.1)	234	272	33	16.2
Billing, cost, and rate clerks (80.7)	216	254	31	17.5

[a] Percent female from 1980 census data (Hunt and Hunt, 1985a:Table 2.4).
[b] Data from Silvestri and Lukasiewicz (1985:Table 2).
[c] Estimates of job growth for women are conservative; because the percentage female is likely to grow by 1995 in many of these occupations, these numbers underestimate job growth for women.

11.5 percent decline). In sum, slow growth is now expected to be even slower, and declines in some occupations will be significant.

One area of anticipated rapid growth for clerical workers is the temporary help industry. Between the trough of the recession in 1982 and December 1985, the number of employees working for firms that supply temporary personnel nearly doubled, making it one of the fastest-growing industries with more than 50,000 workers (Carey and Hazelbaker, 1986). Of the industry's 735,000 employees, more than half are involved in office occupations (other significant areas of temporary personnel include industrial, medical, and engineering occupations). Continued strong growth is expected, but at less than the very rapid rate experienced between 1982 and 1984 when the industry probably benefited from the recovery of the recession; as the recovery continues, employers may be more likely to hire workers on their regular payrolls. The BLS moderate-trend projections estimate annual growth for the industry at 5 percent through 1995, higher than the 4.2 percent annual rate estimated for all business services industries, and much higher than the overall estimated annual increase of 1.2 percent. Employment is expected to be 1,060,000 by 1995 (Carey and Hazelbaker, 1986). The employment figure can be thought of as the average number of daily placements. The number of people working for temporary agencies during a year is much larger; one estimate put it at more than 5 million for 1984 (Appelbaum, 1985). According to Appelbaum's study, the dominant motive of employers in hiring temporary rather than permanent workers is cost cutting: hiring, training, and fringe benefit costs are generally reduced. Appelbaum believes the dominant motive of temporary workers, especially women workers, is to obtain flexible scheduling. In Appelbaum's view there are disadvantages as well as advantages to temporary help, for both employers and employees. Employees receive fewer fringe benefits and are unlikely to experience the earnings growth that normally accompanies seniority with an employer. Employers, Appelbaum believes, may be forgoing the opportunity to restructure work in the most efficient and productive manner in the long run as they opt for short-term cost savings. Whatever the advantages and disadvantages of temporary work, observers agree that employment growth in this sector is likely to remain strong. Given the overall slow growth predicted in clerical occupations generally, the rapid increase predicted in temporary employment constitutes a shift in clerical employment from permanent to temporary work.

Because it is of interest to compare growth in clerical jobs with opportunities elsewhere in the economy, Table 3-21 presents the 20 occupations that are expected to experience the largest growth of jobs by 1995 throughout the entire labor market. In percentage terms their growth rates for the 1984–1995 period vary from 9.6 percent for secretaries and general office clerks to 71.7 percent for computer programmers. A number of the top 20 occupations are related to computers, and several others are related to preparing and serving food, to sales

TABLE 3-21 Occupations with the Largest Job Growth, Bureau of Labor Statistics Projections, 1984–1995

Occupation	Employment (thousands)		Change in employment 1984–1995		Percent of Total Job Growth 1984–1995
	1984	1995	Number (thousands)	Percent	
Cashiers	1,902	2,469	566	29.8	3.6
Registered nurses	1,377	1,829	452	32.8	2.8
Janitors and cleaners, including maid and housekeeping cleaners	2,940	3,383	443	15.1	2.8
Truck drivers	2,484	2,911	428	17.2	2.7
Waiters and waitresses	1,625	2,049	424	26.1	2.7
Wholesale trade sales workers	1,248	1,617	369	29.6	2.3
Nursing aides, orderlies, and attendants	1,204	1,552	348	28.9	2.2
Salespersons, retail	2,732	3,075	343	12.6	2.2
Accountants and auditors	882	1,189	307	34.8	1.9
Teachers, kindergarten and elementary	1,381	1,662	281	20.3	1.9
Secretaries	2,797	3,064	268	9.6	1.7
Computer programmers	341	586	245	71.7	1.5
General office clerks	2,398	2,629	231	9.6	1.4
Food preparation workers, excluding fast food	987	1,205	219	22.1	1.4
Food preparation and service workers, fast food	1,201	1,417	215	17.9	1.4
Computer systems analysts, electronic data processing	308	520	212	68.7	1.3
Electrical and electronics engineers	390	597	206	52.8	1.3
Electrical and electronics technicians and technologists	404	607	202	50.0	1.3
Guards	733	921	188	25.6	1.2
Automotive and motorcycle mechanics	922	1,107	185	20.1	1.2

SOURCE: Office of Technology Assessment (1986: Table 8A-1).

in general, and to health care. Janitors and cleaners, truck drivers, accountants and auditors, and teachers are also included. Clearly the large-growth occupations include both those that require substantial education and training (registered nurses, electrical and electronics engineers) and those that do not (cashiers, fast-food workers). The slowest-growing of these occupations are the clerical occupations.

Job Loss and Displaced Workers

The slower growth projected for clerical workers overall, combined with the anticipated decreases in some clerical subfields and the specific employment losses observed in some instances of automation (see, e.g., Appelbaum, 1984; Gutek and Bikson, 1985), indicate the need for considering programs that will help workers shift occupations as necessary. The relative disappearance of back-office jobs and continued growth in jobs with greater customer contact suggest that workers may benefit from help in identifying their oral communication skills and developing them for transition to jobs with more customer contact.

It is difficult to know how the slower growth overall and the occupational shifts anticipated in clerical work will translate into unemployment, if at all. Some of those who lose specific jobs will remain with their employers; even if many are displaced (laid off more or less permanently), substantial unemployment may not result if those who lose jobs find new ones quickly.

Because of intense interest in the question of displaced workers during the recessions of the early 1980s, BLS added a special supplement to the January 1984 Current Population Survey (Flaim and Sehgal, 1985). Respondents from about 60,000 households were asked whether any adult member of the household had experienced job loss since 1979 because of a plant closing, an employer going out of business, or lack of recall from a layoff. The results show that from 1979 to 1983, nearly 11 million nonagricultural workers lost jobs; this represented about 12 percent of annual employment based on payroll data, a proportion much larger than previous estimates obtained from less complete information (Podgursky, 1986).

Podgursky's analysis of a sample of the displaced workers (those aged 20–61 and displaced no later than December 1982) finds that a substantial proportion found jobs within 15 weeks (about 40 percent) and a substantial proportion remained jobless for more than 52 weeks (about 26 percent). Relative to their representation in the total labor force, the displaced workers were disproportionately blue-collar workers from manufacturing; white-collar and service workers constituted 39 percent of those displaced. Whether white-collar or blue-collar, women remained jobless longer than men; white-collar workers had less joblessness than blue-collar workers, and more women worked in white-collar than in blue-collar jobs. The industries that are especially large employers of clerical workers (for example, government and finance) had much less displacement than others.

An analysis of a somewhat different sample of the displaced workers in the same survey (Flaim and Sehgal, 1985) showed that clerical workers fared somewhat better than average in replacing their former earnings. In general, however, reemployment incurred earnings losses for nearly everyone (average losses for full-time workers were 12 percent in white-collar and 15 percent in blue-collar jobs; Podgursky and Swaim, 1986). Both women and blacks had longer periods of joblessness, and workers with below-average education had larger earnings losses on reemployment. Better-paid and more senior workers faced larger losses, as did those who were reemployed in different occupations or industries. Some 15 to 20 percent of the displaced workers had participated in education or training programs, most often paid for either by themselves or their employers.

CONCLUSION

This review of recent and future trends in labor force growth and clerical employment suggests that technological change is likely to contribute to employment problems for women, but that massive job loss is unlikely to occur. Clerical jobs will experience slower growth in the aggregate than they have in the recent past, and shifts in the demand for various clerical occupations will occur. Some increased structural unemployment may result.

Between 1958 and 1968, clerical employment grew considerably more rapidly than total employment: clerical workers increased their share of total employment from 14.5 to 17.5 percent. Between 1970 and 1980, however, growth was slower: clerical workers increased only from 17.5 to 18.5 percent. And, since 1980, their share of total employment has remained the same. In the "most plausible worst case" scenario developed by the panel—the largest plausible negative impact of information technology on clerical employment, based on the largest historic negative effect in any industry—using the November 1985 BLS projections of slower employment growth overall, clerical employment by 1995 would have lost at most 2 percentage points of its share of total employment. Clerical employment would increase by 2.0 million jobs, or an increase of 10.5 percent, between 1982 and 1995. This growth rate is one-third that of the historic growth rate of clerical employment between 1972 and 1982 (about 0.7 percent per year compared with 2.3 percent per year).

Labor supply is also expected to grow more slowly after 1985, but not as slowly as clerical employment would in the most plausible worst case. Labor force growth will decline to about one-half its growth in the previous decade. Thus, women are expected to enter the labor force somewhat faster than clerical jobs are now expected to grow. In this case, a larger proportion of women than in the past would have to find work in nonclerical occupations. Since women have been integrating formerly male-dominated occupations, the panel does not think such a change would be difficult. In any event, the panel does not foresee massive technologically induced unemployment among clerical workers or would-be clerical workers, and if the economy grows at an average rate, opportunities in other areas should be sufficient for women to shift occupations.

This finding, coupled with the panel's review of the shifts among clerical subfields, supports the need for programs that will assist workers with transition. As noted above, back-office jobs appear to be declining relatively, and in some cases absolutely, while those that require greater contact with customers are increasing relatively. This differential change is likely to pose a particular hardship for minority women, who hold relatively more of the declining jobs. For all women, the slower growth and likely shifts point to the need for sound basic education in core competencies, such as reasoning ability, problem solving, and communication, to prepare workers for the jobs likely to be created. Since technical occupations, such as those related to operating computer equipment, will also grow relatively rapidly, good technical training will also be important.

4

Effects of Technological Change: The Quality of Employment

EMPLOYMENT QUALITY

Early discussions of the impact of technological change focused primarily on the numbers and types of jobs that would be replaced or created (U.S. National Commission on Technology, Automation, and Economic Progress, 1966). Today the debate has expanded to consider the effects of the new technologies—particularly those that affect information—on the quality of employment. Do the new information technologies lead to the fragmentation of work or to its integration? To the deskilling of work or to more highly skilled work? To electronic monitoring of employees or to greater job autonomy? To the exploitation of workers with limited job opportunities or to the freedom to work at convenient times and places? These issues are especially germane to women, because as clerical workers, bookkeepers, nurses, librarians, and other direct users of information technology, they are likely to be affected in large numbers. In addition, their relative lack of power in the workplace suggests that if information technology has pernicious effects, they will bear its brunt. It would be desirable to establish whether information technology is used primarily to increase or decrease the quality of jobs and to determine the conditions under which it does one or the other.

Two images regularly appear in the research literature and public debate on the effects of information technology on employment quality. Some commentators, while acknowledging that technology can be used to improve job quality, believe that it has most often been used to undermine the quality of white-collar work. "Our recent research has, if anything, strengthened our earlier conclusions. More and more evidence . . . documents the deteriorating quality of office work. . . . [T]he introduction of automated office equipment has extended management control over the work process to the detriment of workers' job satisfaction" (9-to-5, 1985:29). In this pessimistic view, information technology is associated with degraded, deskilled, and devalued jobs, stressful and dangerous work, employer monitoring of employees, and work speedups, in which workers are paid less for doing more.

A case from the Project on Connecticut Workers and Technological Change, described by Westin (1985), typifies the pessimistic image. A small, metropolitan-area mail-order jewelry firm, selling low-cost rings, used computers for inventory and automatic billing through an on-line data base. Order division clerks used video display terminals (VDTs) exclusively for data entry and retrieval. The managers monitored workers through automated analysis of their daily computer output and through television cameras focused on their work space. One worker recalled that "they used the cameras to watch how hard you seemed to be working, when you got up to stretch or take a break, and your 'attitude' at work" (Westin, 1985:29–30). Management's objective was to run its business at the lowest cost possible. Both an abundant labor supply and the minimal training needed to operate the equipment and to perform the order-taking job helped keep labor costs low, despite substantial employee turnover. Management succeeded in running a profitable business but at high cost to its workers.

The alternative image portrays information technology as a boon to both employers and employees: increasing workers' productivity, eliminating repetitive or mindless work, providing better tools, and offering intellectual challenge and possibilities for growth. Spinrad (1982) describes his office routine and illustrates with an almost science-fiction-like quality what life in the office of the future could be like. He flips a switch to read his messages on a screen. Communicating via computer mail with a colleague, who electronically forwards an old report, Spinrad then incorporates the report into a message he is writing and sends it to its destination. He then settles into a morning's work of computer-aided hardware design.

Poppel's analysis (1982) of the benefits of office automation for sales and information workers portrays a similar picture. After studying 15 large U.S. organizations, Poppel concluded that a salesperson's time is wasted on traveling, missing contacts, finding out information, and filling out forms, while the time of many managers and professionals is similarly wasted on meetings and clerical work. According to his analysis, office automation technology can rescue some of that wasted time and make jobs more rewarding. For example, he foresees that a salesperson equipped with a portable intelligent display terminal could compute an optimum route for sales contacts, display product and pending order information while at a client's office, calculate cost proposals, and keep in touch with headquarters through electronic mail.

Between the two extreme images of the effects of information technology are analysts who believe that it has had little effect on employment quality except by changing the occupational mix. For example, several years ago Simon (1977) concluded that, in the aggregate, job satisfaction had not decreased over the preceding 20 years, a time of rapid technological change, and that office automation in particular did "not appear to change the nature of work in a fundamental way." Moreover, since office automation consists of labor-saving devices that eliminate jobs that are already relatively routine, he expected to see a decline in the percentage of employees engaged in routine clerical work and a corresponding increase in more satisfying jobs in service, sales, professional, and technical work.[1]

This section asks whether technology, in both its narrow sense of hardware and software and in the broader sense that includes the organizational arrangements through which it is implemented, has increased or decreased employment quality in the aggregate. The focus is on the clerical work generally done by women, although many of the issues are similar for men and women in many kinds of jobs. Most of the available research to answer this question consists of studies of the introduction of technology in particular industries or particular firms. While they provide rich detail about the mechanisms through which technology combines with the organization of work to influence the quality of jobs, they are not systematic samples of either workers or establishments and are thus not helpful in determining which tendency dominates (Attewell and Rule, 1984). In addition, at the present time technology and people's attitudes toward both technology and job design are changing rapidly. These changes are likely to be spreading unevenly across sectors of the economy, further increasing uncertainty about prevailing tendencies.

Defining Employment Quality

Understanding how technology influences employment quality requires criteria for assessing employment quality. But the criteria depend on the perspective from which one defines it. Employers, who may want to optimize produc-

---

[1] Simon's data may have been wrong or outdated. The Quality of Employment Survey (Quinn and Staines, 1977) found no change in overall job satisfaction between 1969 and 1973 but an appreciable drop between 1973 and 1977. The more specific the topic area being queried (e.g., satisfaction with comfort, challenge, financial rewards, or promotion), the larger the decline. Only one question failed to show a drop in job satisfaction: "All in all, how satisfied would you say you are with your job?" Unfortunately, this single item was the typical measure of job satisfaction used in the studies on which Simon based his conclusions (U.S. Department of Labor, 1974). It is not known whether the measured drop in job satisfaction in the 1970s reflected changed working conditions (including but not limited to the effects of automation), changes in expectations about work, or changes in the demography of the labor force.

tivity, product quality, and worker dedication, undoubtedly have different criteria than employees, who may want to optimize intrinsically interesting work, compensation, job security, or pleasant working conditions. Among themselves, both employers and employees may also have different criteria, depending on their social and economic circumstances, group memberships, and individual preferences.

To some degree employment quality is subjective and idiosyncratic and reflects the fit between particular workers' needs and the characteristics of particular jobs. Still, the literatures on job satisfaction, job performance, and job design suggest at least three broad factors that influence employment quality for most workers (Barnowe et al., 1973; Locke, 1976; Hackman and Oldham, 1980): (1) job content; (2) working conditions, especially the social conditions, under which the job is done; and (3) economic considerations.

Job content encompasses those attributes that are intrinsic parts of the work. There are a number of attributes—summarized under the rubric of challenge, especially mental challenge (Locke, 1976)—that increase employment quality. Learning, creativity, autonomy, responsibility, variety, and coping with difficulties exercise workers' conceptual faculties, while the lack of these qualities bores them. Job redesign programs try to infuse jobs with these qualities (e.g., Hackman and Oldham, 1980). For example, if all else is equal, jobs with variety tend to be more rewarding and better for many workers than those in which workers do repetitive work; autonomous jobs in which individuals can pace themselves and control what, how, and when to do their work are better than jobs in which all details are prescribed; integrated jobs in which an individual performs a range of tasks that produce a complete product are better than those in which an individual performs only one fragment of a task or performs a task that produces a fragment of a product. Finally, jobs with feedback, in which workers can evaluate their performance against some standard, are better than ones in which workers perform tasks without knowing how they are doing. When technology is introduced in ways that enhance these qualities, the jobs get better for most workers, as witnessed by an increase in their satisfaction with and commitment to their work (Hackman et al., 1978).

Working conditions are both physical and social. The precise physical conditions that make a job dangerous or uncomfortable vary greatly: they may include limb- or vision-threatening equipment as well as extremes of temperature, noise, and lighting. In introducing technology into white-collar work, workstation and office design have been major concerns, ranging from fears of permanent vision damage and problem pregnancies to discomfort with seating position and vision (National Research Council, 1983; Westin et al., 1985).

Social conditions that give workers satisfaction include being with other people on the job. On-the-job friendships are especially important sources of social satisfaction and support for white-collar workers because communication is often an integral part of their jobs (Panko, 1984). Social satisfaction and support from coworkers are among the reasons that people who are employed have better mental health than those who are not (e.g., Thoite, 1983). To the extent that using new technology isolates workers, keeping them at terminals or at home away from others, it may disrupt this source of job satisfaction. On the other hand, like the telephone, it also opens up new occasions for and ways to communicate with others.

Another aspect of social conditions on the job is supervision. Supervision can be more or less close, more or less confining, and more or less helpful—with or without new information technologies. The new technologies do, however, facilitate more detailed electronic monitoring—of keystrokes per minute, for example—and may therefore contribute to closer, more onerous supervision. As noted above, autonomy and control over the pacing and methods of work are an important factor in job quality.

Economic considerations include both the absolute level of wages and salary, fringe benefits, security, and promotion possibilities that a job offers, and the fairness of these factors compared with the norms of an occupation and industry and with the inputs, such as seniority, education, and skill, that a worker provides. As technology changes the occupational mix in an industry and the skills demanded by particular occupations, it is likely to have a direct effect on workers' perceptions of their job security. In addition, workers often believe that they should be compensated for the new skills they have acquired in using the new technology (Murphree, 1985).

How is information technology associated with these sources of employment quality? This section attempts an answer first by trying to assess workers' satisfaction with information technology and the jobs that use it and then by examining the specific factors noted above: job content, especially job fragmentation and deskilling; working conditions, such as computer monitoring and work pacing, telecommuting and the electronic distribution of work, and ergonomic conditions (the interaction of equipment features and its use by humans); and economic considerations.

Workers' Satisfaction and Attitudes

There is no research surveying a representative sample of workers in the United States about the technology they use to do their jobs and their perceptions of the quality of their jobs. One of the most thorough sources of information about employment quality, the University of Michigan's Quality of Employment Survey, conducted in 1969, 1973, and 1977, provides little information on technology, and in 1977 it explicitly dropped questions about the effects of automation (Quinn and Staines, 1977). There have been several large-scale surveys, but one should be cautious in generalizing from them because of likely biases in their sampling procedures, their sketchy detail about how and how much technology is used in jobs, the retrospective nature of the questions they asked, and the vested interests of some of the sponsoring organizations. These surveys have been commissioned or conducted by manufacturers, employers, labor groups, and academic researchers. For example, the Honeywell Corporation, a vendor of office automation equipment, commissioned a survey comparing the reactions to office automation of 937 managers and 1,264 secretaries in a national random sample of 443 establishments in information-intensive industries. The Minolta Corporation, a copier manufacturer, and Professional Secretaries International, a worker organization, commissioned a similar survey of more than 2,000 secretaries and their managers in 22 cities. Kelly Services, a supplier of temporary clerical and other workers for business, commissioned a survey of 507 secretaries in more than 500 establishments, and 9-to-5, the National Association of Working Women, a labor group, conducted a large-scale questionnaire survey focusing on job stress among women. Kling (1978) surveyed 1,200 managers, data analysts, and clerks in 42 municipal governments. Bikson and Gutek surveyed and interviewed managers, professionals, and clerical workers in 26 manufacturing and service organizations in California (Bikson and Gutek, 1983; Gutek et al., 1984; Bikson, 1986). Westin and his colleagues visited 110 business, government, and nonprofit establishments, conducting more than 1,100 open-ended interviews with end-users of VDTs, primarily at the clerical, secretarial, and professional levels, and with more than 650 managers and executives (Westin, 1985; Westin et al., 1985).

The major problem with most of these surveys is that their samples are not representative; and in no case did a survey provide a representative sample of a broad range of users of information technology. More importantly, because the researchers were haphazard in their respondent selection, because managers selected the workers to be interviewed, or because workers selected themselves, in most cases the reader does not know to whom the conclusions apply. For example, the Honeywell survey used a representative sampling of business establishments but not of the employees within the establishments: the chief executive's secretary provided the names of secretaries and managers to be queried; the survey was limited to secretaries working for one or more managers; and secretaries who worked in secretarial pools were excluded. Uncontrolled biases in these selections directly influence the portrait of office automation that results from the survey. The Kelly Services survey had similar selection biases and restrictions.

The 9-to-5 survey on women and stress is based on a self-selected sample of women who responded to a questionnaire printed in four monthly women's magazines. The ways in which these respondents differ from other working women who neither read the target magazines nor responded to the survey are unknown.

The Bikson and Gutek survey and the Westin survey used opportunity samples—the researchers talked to people in organizations in which they could most conveniently make the necessary contacts and get the necessary permissions. These organizations may be very different from less convenient or less forthcoming ones. In addition, Westin concentrated on visiting organizations with "reputations as 'advanced' and 'active' users of office systems technology . . . [with] 'good human resource policies'" (Westin, 1985:3).

The second problem with these surveys is that the questionnaires may have introduced distortions because of the way some questions were asked. A number of the surveys asked the same respondent to compare work before and after the introduction of information technology and to attribute job-enhancing or job-degrading effects to the technology (Kling, 1978; Honeywell Corporation, 1983; Minolta Corporation, 1983; Kelly Services, 1984; 9-to-5, 1984a). The problem is that people often have difficulty remembering attitudes and atmosphere from the past, even the recent past, and have even greater difficulty accurately attributing causes to the changes they perceive taking place (Bem and McConnell, 1970; Nisbett and Wilson, 1977). Respondents are likely to reconstruct the past on the basis of their current environment and their theories of the impact of information technology, so workers may believe, for example: "I like my job now and everyone knows that word processors make jobs better, therefore before the technology my job must have been worse," or, "I have occasional headaches and everyone knows that staring at a screen makes headaches worse, so I must have more now than I did before I got my terminal."

In addition, the phrasing of some questions may have biased respondents' answers. For example, the phrasing of questions in the Honeywell survey probably had the effect of portraying office automation equipment in a positive light: respondents were asked to agree or disagree with statements that automated equipment made tasks easier, made tasks faster, improved work flow, made jobs more challenging, and improved the quality of work. But respondents were not given similar opportunities to indicate that automated equipment made jobs less interesting, increased stress, or made skills obsolete.

This direct bias in question design, however, was surprisingly uncommon given the vested interests of some survey sponsors. For example, the Kelly Services survey (1984) allowed respondents to indicate that they loved or hated word-processing equipment, that information technology made their jobs more or less stressful, and that they have had stress-related physical symptoms. Similarly, the 9-to-5 (1984a) survey allowed respondents to indicate that they were treated with respect or hostility by their managers, that the introduction of automated equipment decreased or increased job stress, or that they had great or little job autonomy.

The third problem with these surveys is that many factors that both influence the quality of employment and are associated with the use of information technology have not been adequately controlled in data analysis. Thus, such factors as the industry in which employees work, the size of the establishment in which they work, their occupation, their seniority, and their age are rarely controlled for when examining the impact of information technology. In the insurance industry, for example, if it is true that a data-entry clerk has a worse job than does a claims adjuster and that, on average, a clerk uses information technology more than an adjuster does, an analysis that inadequately controls for occupation might mistakenly conclude that use of information technology rather than other aspects of an occupation is associated with poor jobs.

The fourth problem arises because large-scale surveys must generalize across different industries, occupations, technologies, and work arrangements; consequently, they lose qualitative detail about the processes through which information technology is introduced into the workplace and the effects it has. (This point is elaborated below.)

Despite these limitations, however, these surveys provide a valuable source of information about the range of conditions under which information technology is introduced and the distribution of its effects. If they are considered with these methodological reservations in mind, they can provide suggestions about general tendencies that are not derivable from the case-study literature. Overall, these surveys suggest that workers who use information technology are generally satisfied with it, because it allows them to do their work better and because it improves the jobs themselves or, at a minimum, does not degrade them significantly. For example, the Honeywell survey found that both managers and secretaries liked the technology, agreeing that it improved the quality of work, made routine tasks go faster, allowed more to get done each day, and freed time for more interesting and challenging work (Honeywell Corporation, 1983). The survey by the Minolta Corporation and Professional Secretaries International reported similar results: about 90 percent of the secretaries believed that office automation had made them more efficient and productive, and about 70 percent believed that it made secretarial jobs more fulfilling, providing more time to do challenging and interesting work (Minolta Corporation, 1983). Bikson and Gutek (Bikson and Gutek, 1983; Gutek et al., 1984; Gutek and Bikson, 1985; Bikson, 1986) and Westin and his colleagues (Westin, 1985; Westin et al., 1985) report similar results. The survey of women and stress by 9-to-5 (1984a) collected extensive information from 5,000 women workers. The majority of the respondents reported that their jobs were more interesting and enjoyable after automation (68 percent) and less stressful and pressured (54 percent); these perceived benefits of office technology were stronger for managerial and professional workers than they were for clerical workers. Health data, however, contrast with these attitudinal data: users of VDTs and microcomputers reported more frequent physical and psychological symptoms, such as eye strain, chest pain, tension, depression, and vision problems requiring a doctor's consultation, than did nonusers.

Kling's (1978) survey covered workers at different levels in municipal governments who used computer-based reports, not on-line VDTs. The respondents attributed broader job content to computer use. Respondents also attributed increases in job pressure, but not closeness of supervision, to computer use. These effects increased as the respondents used computers more. As in the 9-to-5 survey, Kling found that computers enhanced the jobs of higher-level workers most. Kraemer and Danziger (1982) also analyzed survey data from the same large sample of municipal government employees and found that about half the workers experienced an increased sense of accomplishment from computerized work, while few experienced a decreased sense of accomplishment.

One may distrust these studies and their generally sanguine conclusions because of the level of detail that such surveys achieve as well as for the methodological reservations listed above. By necessity, these studies asked questions at a very abstract level to generalize across different industries, occupations, technologies, and work arrangements. As a result, they have undoubtedly missed some of the behavioral changes in work caused by technological change and many of the mechanisms through which these changes occur.

The changes in work caused by technology can be subtle, unanticipated, and difficult to derive from any general theory of technological change or any survey of job satisfaction. For example, fears about the introduction of electronic cash registers in retail work center around the deskilling of sales workers, work speedup, and the continued growth of part-time, dead-end employment. Irrespective of the validity of these fears, other consequences of electronic cash registers were not anticipated. A cash register's capacity to compute change determines the procedure cashiers use to return change to customers (and, as a likely consequence, their facility with mental arithmetic): with mechanical cash registers, cashiers often use a subtraction by addition rule, returning coins until an even dollar amount is reached and then returning bills; when the register computes the change, clerks return bills first (Fleming, 1985). Some of these subtle changes have substantial impact on the quality of jobs for cashiers. On mechanical registers, cashiers often key in all the items for one customer and then turn to bag them in a separate operation. With faster electronic registers, a "ring-and-bag" process is often used—each item is keyed in and bagged immediately. This process, which is faster since each item is handled only once, leads to pain and health problems: cashiers stand off balance and bag goods with only the left hand, putting disproportionate strain on the left side of the body. The more extensive introduction of electronic scanners at the checkout counter may relieve these health problems, because the goods can be pulled over the scanner and bagged with two hands (Ontario Retail Council, 1981; Wallersteiner, 1981; White, 1985). Zuboff (1982) offers additional observations on the subtle effects of information technology.

Aggregate studies may also generate misleading conclusions about the mechanisms by which technological effects occur. For example, commentators assume that if information technology increases management control of workers, it does so because managers use machines to gather more detailed information about their subordinates (9-to-5, 1985). However, Kling and Iacono (1984) report that the need for accurate data to generate useful computerized reports redefines work activities, and subjects workers, regardless of their position in the corporate hierarchy, to tighter social controls. The many groups needing accurate information put pressure on those who generate it to work according to standard procedures (cf. Baran, 1985).

When commentators discuss the changes in skill requirements associated with information technology, they often assume either that the technology subsumes substantive knowledge of a job and its procedures (e.g., Baran, 1985) or that it reintegrates jobs that had been previously fragmented for other reasons (e.g., Giuliano, 1982). But mastering the information technology may itself require skill, especially if the technology is powerful, complex, defective, or an imperfect fit to local conditions. For example, in his case study of the insurance industry, Attewell (1985) describes the emergence of an informal class of computer gurus or mavens—noncomputer professionals who become local experts in beating the system, in circumventing its bugs, and in coercing it to respond appropriately to nonstandard local conditions. Managers provided these gurus with "nonproductive" time on the computer to learn its idiosyncracies, and they, in turn, conducted informal seminars, translated headquarters' instructions into understandable procedures, and invented procedures for getting around computer intransigence. As a result of their expertise, they gained substantial job satisfaction and sense of mastery and skill and substantial informal power in the organization.

JOB CONTENT: JOB FRAGMENTATION AND THE DESKILLING DEBATE

Braverman's seminal work (1974) has shaped the terms of the debate about the effects of information technology on the content of jobs. He argued that the introduction of technology into the office is a mechanism to rationalize and fragment office work. As in a factory, office work is broken into many subtasks, each performed by a "detail" worker, who loses the integrative contact with the total product and who loses variety in the job. The consequences from the employer's perspective are to reduce the skill requirements of office work— that is, the average employment experience and education needed to adequately perform the job—and therefore to reduce labor costs. The consequences from the employee's perspective are to reduce the quality of the job by reducing the variety of tasks performed, the mental challenge, and job autonomy and responsibility.

Studies have documented many episodes in which the introduction of new information technology coincided with a fragmentation of jobs and a decrease in the skill levels required to do them. For example, Murphree (1984) describes the fragmentation of the job of legal secretary in the large firm she studied. The fragmentation was not necessarily caused by the new technology, but it facilitated an ongoing process of routinizing work. The variety of tasks that made the job of legal secretary both interesting and highly paid were assigned to other workers: paralegal aides composed first drafts of simple contracts, specialized legal librarians searched for legal citations, and messenger services hand-delivered urgent documents. With word-processing equipment, even the task of producing a manuscript was divided into an initial data-entry stage and a later proofreading and editing one, with different personnel performing each function. Relieved of many of their duties, the legal secretaries found themselves doubled up, working for more than one lawyer, primarily in a gatekeeping role. In this case, new technology was a concomitant of decreased variety, less challenge, and less responsibility.

Researchers have also identified cases in which information technology incorporated substantive knowledge of a job and its procedures, an establishment, or an industry, leaving less for a worker to know. In the insurance industry, the skilled work of assigning risks or assessing claims has increasingly been codified into computer software, so that less skilled, less experienced, and less educated clerks can perform the work once performed by skilled clerks and professionals (Baran, 1985). In social science, sophisticated statistical analyses that once were done only by professionals are now performed by undergraduates or research assistants using a statistical analysis computer program. In supermarkets, knowledge of brands, in-store promotions, and arithmetic is less necessary to a checkout clerk using a bar-code reader and an intelligent cash register: for example, when produce is coded, the clerk need not know the difference between root parsley and horseradish. In these cases the uses of information technology have led to the creation of more routine jobs.

In contrast to these cases in which technology is a tool to restructure work and fragment jobs, other researchers have argued that new office technologies can reintegrate jobs that had been previously fragmented for other reasons (Matteis, 1979; Strassman, 1980; Giuliano, 1982) and require more skill and responsibility of workers (e.g., Attewell, 1985; Baran, 1985). At the clerical level, for example, by using centralized data bases, customer service representatives can handle all of the transactions associated with a client's account, taking orders, entering data, making adjustments, and answering inquiries (Baran and Teegarden, 1984; Baran, 1985; Feldberg, 1986). At a more professional level, using new workstations managers can have greater control over more stages of their work through computer-based tools for searching corporate data bases, calculating the consequences of investment strategies, creating illustrations, outlining, writing, checking spelling, formatting, and sending a finished report to a mailing list of recipients (Spinrad, 1982).

Moreover, starting from the assumption that the most routinized jobs in an industry are the first to be automated, some analysts have argued (Adler, 1984a; Baran, 1985) that automation increases skill requirements for the remaining jobs, especially generalized cognitive and problem-solving skills; increases worker responsibility; and increases coworker cooperation. For example, if most routine banking operations are performed by customers themselves using automatic teller machines, the banking tasks remaining to human tellers are nonroutine, e.g., handling problem inquiries that may require considerable knowledge of banking procedures, problem-solving skills, and skill at dealing with people. Similarly, Baran (1985) notes that, as some work by insurance professionals is being deskilled and redistributed to clerks and machines, the remaining professionals function as exception handlers and do more skilled work. Attewell (1985) also notes that computerization is eliminating the routine work of insurance examiners, such as calculation of deductibles or identification of potential duplicate payment, while at the same time leaving examiners more time to make decisions about dubious claims. Similarly, social scientists using modern statistical software are relieved of the tedium of calculation and can undertake more challenging intellectual tasks.

From these studies it is clear that both job fragmentation and deskilling and job reintegration and upgrading are occurring, but it is not clear which trend is predominant. Aggregate data, although flawed, show little evidence of wide-scale deskilling either within particular industries or in the labor market as a whole and, indeed, show some evidence of increased skill requirements (Attewell and Rule, 1984). Like most of the research reviewed in this chapter, the data are suspect, and the conclusions based on them should be viewed as tentative. The wide variety of skills needed in the workplace and their uneven distribution across jobs and industries and over time makes comparison of skill levels exceedingly difficult. To assess the skill requirements of jobs, researchers have often resorted to the use of one or two proxy measures, such as years of experience or years of education of workers. These measures lose detailed qualitative and quantitative information about skill differences, and they may reflect changing tastes of employers or demographic shifts in the population unrelated to skill requirements of jobs (Rumberger, 1984).

Researchers using assessments of the skill requirement of jobs from the *Dictionary of Occupational Titles* (DOT) and similar sources have found little evidence for the deskilling of white-collar work, either within particular occupations or, more broadly, across occupations. For example, Attewell (in press), using the Bureau of Labor Statistics' skill-level categorizations for 13 occupations within the insurance industry, found that between 1966 and 1980 (a period of intense automation) 4 showed statistically significant downgrading, 6 showed significant upgrading, and 3 showed no trends. He also found a substantial increase overall in the proportion of the insurance work force in higher-level white-collar occupations (from 38 to 60 percent). Using DOT data, researchers have found little change in the skill requirements of a sample of industrial and service occupations from 1949 to 1965 (Horowitz and Herrenstadt, 1966) or more broadly from 1965 to 1977 (Spenner, 1979) and either no changes in the aggregate skill of the nonfarm sector from 1900 to 1970 (Dubnoff, 1978) or a small aggregate upgrading of skills from 1960 to 1976 (Rumberger, 1981).

To the extent that some automated jobs indeed require less skill than nonautomated ones, one still needs to ask about the role of technology in these changes. Scholars are now debating whether new technology encourages job routinization and fragmentation, is a beneficiary of it, mitigates it, or is independent of it. Certainly, scientific management of office work was occurring well before the current introduction of microelectronics (see Leffingwell, 1925; M. Davies, 1982; Murolo, 1986; Chapter 2, this volume). Some have argued that job fragmentation and routinization result from the inherent conflicts of interest between capitalists and workers (Braverman, 1974; Driscoll, 1980), and others have argued that it is a bureaucratic response to increases in office size and paperwork resulting from the increased size of firms, their movement into national markets, or government record-keeping requirements (Murphree, 1984; Strom, 1985). Finally, others argue that fragmentation and routinization are necessary preconditions for the application of the simplistic computer technologies that are available at the current stage in the automation of office work but are not necessarily consequences of the automation (Baran and Teegarden, 1984). According to this view, work that has already been simplified is a prime candidate for automation.

Stages of Technology

What causes the contradictory impressions conveyed by the literature on job fragmentation and deskilling? In part the disagreements are the result of comparisons of experiences at different stages in the evolution of the technology, variations in the uses of the equipment, and differing social and economic circumstances under which new technology is introduced. Given rapid changes in the technology itself, in its uses, and in the cumulation of its effects, conclusions from one wave of technology use may not generalize to later waves.

An examination of the social organization of word processing is an instructive case. Large pools of workers are much more likely to be the modal organization of personnel if the word-processing equipment is old (pre-1978) rather than new (post-1981) and if the firm introducing the technology is large (Benoit et al., 1984). Women who work in large pools have more homogeneous and boring work than do other word-processing workers. They spend more than 90 percent of their time doing typing tasks, while those who serve several clients or who work as private secretaries for a single principal have a much greater variety of tasks (Benoit et al., 1984; Panko, 1984).

It is not clear whether these differences in the social organization of word processing reflect the differing technical capabilities of old rather than new equipment (i.e., minicomputer-based rather than microcomputer-based word processors) or changes in management philosophy resulting from experiences with previous implementations (i.e., a "learning curve") or other social influences. But the point remains that researchers trying to understand the role of technology in job fragmentation would have come to different conclusions had they studied large or small firms using old or new equipment.

Levels of Analysis

A second cause of the disagreement in the research literature results from ignoring important level-of-analysis distinctions among the units of work that are being skilled or deskilled. The skills and skill requirements of tasks, individuals, jobs, occupations, firms, industries, and the labor force as a whole can change with the introduction of technology, but a change at one level has no necessary implication for changes at other levels. A particular task may require less skill to perform when it is done on a computer, but depending on the variety of tasks comprising the job and the knowledge of the worker in that job, the job itself may become more satisfying and challenging.[2] This paradox may be compounded as tasks migrate between jobs and occupations and as new workers fill the transformed jobs.

For example, consider the use of a word-processing system. A standard word-processing system today allows text editing of manuscripts, formatting (e.g., centering, underlining, justification, pagination, and setting fonts and point sizes), sorting, and syntax and spelling checks. As a direct result of the word-processing system, the skill and educational requirements needed to perform some specified tasks have decreased: to sort the entries in a bibliography, a secretary needs less facility with the alphabet; to proofread a manuscript, less

---

[2]This discussion proceeds as if the skill and education needed to perform a job were solely functions of the tasks comprising the job. Clearly, however, the skill requirements of a job have political and social as well as technical foundations. For example, the increased educational requirements for secretaries found in employment advertisements over the last 30 years may reflect the supply of college-educated women, the association of education with other "desirable" class and racial characteristics, or the quality of contemporary education more than they reflect changes in the secretarial occupation.

knowledge of spelling and grammar. In addition, some of the procedures needed to achieve a certain result have changed (e.g., centering or underlining text), although it cannot be determined in advance whether the change in procedure represents an increase or decrease in skill requirements. Finally, some tasks in creating a manuscript electronically have no counterpart in a typewritten manuscript: for example, setting the size and style of fonts is more similar to traditional production editing jobs than to typewriting jobs. It is not clear in the abstract what the net effect of these changes in tasks will be on the skill required to produce a manuscript. Even if, overall, less training, skill, and education are needed to create a manuscript using a word-processing system than using a typewriter, there is no necessary connection between this decreased skill requirement and the skill exercised by particular typists or secretaries, the skill requirements of the job of secretary within a particular establishment, or the skill or educational requirements of the secretarial occupation as defined in private and public occupational classifications.

The skill requirement of jobs depends on the way their tasks are organized. Some jobs may be composed only of tasks that have been clearly deskilled when new technology was introduced: for example, the typing, proofreading, and correcting functions have been separated, and a skilled job is replaced by several new ones, some of which clearly require less skill. In contrast, a secretarial job can be organized, as it is in many universities, so that secretaries perform many services for the principals for whom they work. In this case, the increased speed at performing some of the tasks may lead to more time and energy available for other tasks, including, in the case of university secretaries, advising students, monitoring grants and contracts, and maintaining records. This kind of change is likely to lead to skill enhancement for the job as a whole.

Again, the amount of skilling or deskilling of tasks and jobs has no necessary implication for the skills that a particular job incumbent can use. Clearly, as new technology is introduced, work procedures will change, and some of a worker's knowledge will become obsolete while new knowledge will have to be acquired. But knowledge change is not equivalent to deskilling. In the case of an individual, whether a deskilled job is one that is not challenging because it does not require the individual's mental or physical capabilities is a function of both the skill level required by the job and the skill level brought to it by the job holder. When the same people hold jobs before and after the introduction of technology that lowers the skill requirements of those jobs, they are unable to use their knowledge, skills, and experience, and their jobs become less challenging and satisfying (see, e.g., Rogers and Friedman, 1980). But the introduction of technology that lowers a job's skill requirements often affects new labor force entrants, rather than incumbent workers, who may retire or be promoted, transferred, or laid off (see, e.g., Rogers and Friedman, 1980). The new entrants—often with less employment experience and education—may well be challenged by the level of work. Indeed, the deskilling of some white-collar jobs may be the vehicle by which less advantaged social groups gain white-collar work. And for the same reason, less advantaged workers may face greater job insecurity because those jobs may be at risk of further deskilling or elimination through new technology.

In addition to changing the skill level required to perform certain tasks, the introduction of new technology is often the occasion for the reorganization of existing tasks across job categories. In the law firm that Murphree (1984) studied, for example, the shifting of duties to specialized personnel (e.g., legal librarians using data-base computers) was one element in the deskilling of secretaries' jobs. In universities many professors do much of their own typing on personal computers, taking on data-entry tasks from secretaries, while at the same time achieving more control over the form and content of their writing. In the insurance industry, some of the less skilled tasks that examiners once performed have been transferred to data-processing clerks, upgrading the jobs of both the examiners and the clerks (Attewell, 1985). When tasks are transferred between occupations, shifts in the skill requirements of one job may be associated with complementary shifts in the skill requirements of another.

Occupations are the aggregation of similar jobs across the economy. Determining whether an occupation has been deskilled entails problems similar to those in determining whether a job (rather than a task) has been deskilled. In addition, one must contend with shifts in the definition of the occupations. In particular, as noted in Chapter 3, the Census Bureau and the Bureau of Labor Statistics periodically adopt new occupational classifications. These changes have made all comparisons across time, for both the numbers of workers and quality of work, extremely difficult.

Finally, in asking whether the skills of the labor force as a whole have been affected by technology, one must remember that occupations and industries that require many or few skills can grow or shrink independently of changes in skill requirements for particular occupations. Thus, as Simon (1977) predicted, the skill requirements of the labor force as a whole could rise if low-skill jobs die out. Of course, the reverse could also happen. In this regard it is interesting to note that the greatest new job growth is expected to be in fairly low-skilled occupations, such as food handler and janitor.

Conclusion

Examples abound of both positive and negative effects on skills and job quality when new information technologies are introduced. Using the research literature to ascertain which tendency prevails is difficult because of the different, coexisting stages of technology, the variety of uses of technology, and the differing social and economic circumstances of the workplaces studied. Even the same technology is used in different ways in different workplaces. No study of technological change, the quality of employment, or the effects on skills is based on a representative sample of workplaces and technologies in use. Moreover, information technology and its use are currently changing rapidly; even a well-designed study might miss important new effects on skills.

The available research studies also differ in their levels of analysis. Changes in tasks that occur with new technological capabilities do not necessarily translate into similar changes at the level of the job (or the labor force as a whole). As noted, easier word entry could make a typing job more boring and repetitive or it could make it more challenging if time is freed for other tasks. Thus, the available research literature and the methodological difficulties do not allow the panel to conclude whether there has been an increase or a decrease in the skills required or the quality of jobs with the use of new information technologies. And whatever the overall net effect technology has on skill requirements, there will be some jobs that require less skill and experience after new technology is introduced while other jobs will require more. Consequently, the changing skill requirements associated with technology, regardless of the direction of change, are bound to produce gaps between the skills that job incumbents and entrants have and the skills that jobs require. These gaps are a problem that require attention on their own.

Working Conditions

Monitoring and Work Pacing

The new information technology increases the amount of evaluative information that managers collect and analyze about their workers; such information can enhance the detail, comprehensiveness, and speed of organizational control systems (Kling, 1980). Often this information has been used to assess the amount of work employees do and thus place pressures on workers to meet production standards. Clearly the new information technology can be used to monitor workers more closely, and the literature documents numerous cases where it is so used. The mail-order company described at the beginning of this chapter used both computer-generated reports and television surveillance. The U.S. General Accounting Office installed a computerized security system that was later used to monitor the arrival and departure times of its white-collar employees. The U.S. Army has used computer surveillance to monitor both work time and productivity of individual computer programmers (McDavid, 1985). Word-processing hardware and software often come with report capabilities for monitoring work load and "in the worst sites, centralized [word processing] . . . is designed as an industrial assembly line, emphasizing line counts and time spent on line" (Rice et al., 1983:8).

Whether to monitor the productivity of workers on an individual level is clearly a social choice. Even though they have the technical capability for such monitoring, many organizations do not collect such data on individual workers. Johnson and her colleagues (Johnson et al., n.d.) report that among the 200 word-processing installations they surveyed in the early 1980s, 37 percent measured individual operator performance, and another 8 percent collected performance measures, but used them for planning and group evaluation rather than for individual employee evaluation. When individual measures were used, these were approximately equally spread between measures of quantity (line, pages, or document counts), measures of timeliness (turnaround time), and measures of author or user satisfaction. Thus, only 12 percent of the organizations surveyed collected and used quantitative measures of output as a basis for individual evaluation. Of course, monitoring productivity on either an individual or a group basis may have positive as well as negative effects. Members of work groups might use work-load information to help each other with peak loads. Monitoring individuals might substitute information for prejudice in performance evaluation.

As in the case of deskilling, the data on the overall extent of computerized monitoring are both sketchy and contradictory. Kling's (1978) study of municipal governments found that those workers who used information systems reported increased influence over others, but no overall increase in surveillance and rare monitoring of subordinates' work activities through computerized reports. Kling inferred that employees and managers used the computerized information systems to gain influence in negotiating with their peers and clients. But the 9-to-5 (1984a:4) survey of women and stress found that about 17 percent of women who use computers or VDTs on their jobs report that their work is "measured, monitored, 'constantly watched,' or 'controlled,' by machine or computer system." This effect was larger for clerical workers (20 percent) than for managerial and professional workers (14 percent). Those who reported computerized work monitoring had higher frequencies of a number of stress-related physical and psychological symptoms, including headaches, nausea and dizziness, digestive problems, chest pains, and depression. Furthermore, compared with women in nonautomated offices, women who used a computer or VDT at work were more likely to report that they were required to complete a certain amount of work per hour or day; this effect was also substantially larger for clerical workers than for managerial workers. Again, workers who were subjected to production quotas were much more likely to rate their jobs as very stressful, to report a number of stress-related symptoms and medical conditions, and to have missed work time due to health problems.

Telecommuting and the Electronic Distribution of Work

"Telework," remote work, or telecommuting is the use of computers and telecommunications equipment to do office work from homes or other locations away from a conventional, centralized office. The increasing numbers of women in the labor force with young children, the decreasing costs of computer and telecommunications equipment and services, the increasing amounts of information available in electronic form, and the increasing proportion of the work force performing information work are all trends consistent with increased teleworking (see Applebaum, 1985; Kraut, in press; and Kraut and Grambsch, 1985, for alternate views). Telework is a work alternative that some people have claimed is especially appropriate for women who must combine family and paid employment responsibilities. In a large sample of home-based workers, Christensen (1985) notes that women with children often decided to remain at home with their children and used home-based work as a mechanism to uphold traditional family values while at the same time earning income or maintaining career paths.

A sharp debate exists between those who see women using telework as an alternative work style to combine paid employment with family and child care responsibilities (Olson, 1983; Pratt, 1984) and those who see electronic homework as continuing the traditional exploitation of isolated, predominantly female, home-based workers (Chamot and Zalusky, 1985). The latter observers fear that protective labor laws, which were instituted to curb child labor and other sweatshop abuses, are more difficult to enforce when employees work at home. For this reason, the AFL-CIO and 9-to-5 have called on the Department of Labor to institute a ban on teleworking for clerical workers.

Because so little telecommuting exists, it is impossible to get convincing evidence on this issue. According to the 1980 census, only 1.6 percent of nonfarm workers worked at or from home on their primary job, a number that has been declining since 1960 (Kraut and Grambsch, 1985); undoubtedly, a much smaller percentage has been working at home using computers and telecommunication technology. However, the census is likely to underestimate somewhat the numbers of home-based workers: respondents are asked about the work location of only their primary job and their primary workplace, so the number excludes people who moonlight from home on a second job, who supplement office-based work with home-based work, or who work from home only occasionally. Whatever the overall prevalence, there are concentrated pockets of home-based workers who are likely to use information technology on their jobs, including owners and employees of typing and word-processing services and copy editors, indexers, and proofreaders in the book publishing industry.

One can draw some implications about telework from the limited number of corporate pilot projects on telework and from surveys of home-based workers more generally. Telework pilot projects typically allow a small number of volunteer employees in large corporations to perform most or all of their normal work from home (see Olson, 1983, 1986; Pratt, 1984; Board on Telecommunications and Computer Applications, 1985). These projects probably overrepresent socially acceptable pilot projects from socially responsible employers, who are also the ones most likely to accept researchers. These projects have generally found that employees enjoy telework, that teleworkers are productive and can be supervised, and that a dominant motivation for employees to work at home is to mesh their working lives with their personal lives. For men, it often means working at places and times that are convenient; for women, it often means integrating paid employment with family responsibilities. As Pratt (1984:12) noted in her interview study of 46 teleworkers:

They wanted 24 hours in which to integrate their work and personal lives. By preference, they used daylight time to swim, play golf, or walk on the beach. Parents balanced their work and family responsibilities; for example, a father supervised two children so the wife could leave the house. Mothers put the clothes in the washer, the children down for naps, and then sat at the computer terminal to work.

When evaluations of pilot projects have examined conditions of work, they have found that home-based professionals (primarily men in these samples) retain their salary and benefits and job security, while home-based clerical workers (exclusively women in these samples) experience decreased pay, benefits, and job security (e.g., Olson and Primps, 1984; Olson, 1986). In particular, home-based clerical workers are often hired as contract employees and are paid on a piece-rate basis.

While the current controversy has focused on telework, one can illuminate it and place it in context by examining home-based work more generally. Indeed, many policy issues surrounding home-based work are identical regardless of the tools used to do the work; for many questions it is irrelevant whether workers use regular mail or electronic mail, voice telephone or data transmission, or typewriters or personal computers.

Kraut and Grambsch (1985) analyzed the demographic and economic situation of home-based workers as identified in the 1980 census. Their results suggest that homework is a work-style arrangement that people use to gain flexibility in employment, but that flexibility is gained at the price of lost income. In particular, women with young children are overrepresented among home-based workers, but only if they are married. The nature of gender roles in the United States means that women with children, especially young children, have exceptional constraints on their time, which are not shared by men. They often require increased employment flexibility to handle the dual demands of child care and paid employment. While unmarried women with children also need employment flexibility, they need money more and cannot afford the low pay associated with home-based employment. Other needs for flexibility also lead people to work from home; as a result, the elderly, the disabled, and people living in rural areas are overrepresented among home-based workers.

Home-based workers earn less than conventional workers, in part because they are far more likely to work part time or part of the year. But even among those who work full time, year round, workers at home earn only 76 percent as much as workers in conventional locations, even after controlling for self-employment, occupation, and many of the demographic and background characteristics that vary with income.

The causal link by which home-based workers earn less than conventional workers still needs to be established. Christensen (1985) notes that both clerical and professional home workers frequently work on a contract basis for a single employer. The employer, however, typically does not treat them as employees, providing no fringe benefits or guaranteed work. Historical comparisons suggest that home working is a mechanism by which employers pay those with few labor market alternatives less than they do other workers (Daniels, 1984). Kraut and Grambsch (1985), however, did not find that home-based working decreased earnings most for the most vulnerable groups; they found that it had a uniformly depressing effect on earnings, being the same for women and men and for all women regardless of family status or other demographic characteristics. While the effects of home working differed for different occupations, clerical and other relatively low-paying occupations with large proportions of women were not differentially affected.

Taken together, analysis of corporate pilot projects and analysis of census data suggest that telework, if it flourishes, will be used like part-time work and contract work (Applebaum, 1985) to provide flexibility for some women to meet two sets of demands: paid employment and household responsibility. This flexibility, however, is gained at a price, since people who work at home are likely to earn substantially less than those who work in conventional locations. Because telework may be increasing through the use of technology and because, historically, home workers and their families have been exploited, the extent of telework and the conditions of those who work at home need to be carefully studied.

Ergonomics: The Fit Between People and Technology

A major difference between the last wave of white-collar automation in the 1960s and early 1970s and the current wave is the style with which workers communicate with the new equipment. Gone are operators typing data and instructions on cardboard cards and receiving results hours later. One of the characteristics of the new information technologies is that many workers, both clerical and professional, spend much of their work time sitting in front of a VDT. This work style has been coupled with numerous complaints about visual system impairment, vision fatigue, muscular discomfort, and pregnancy problems (9-to-5, 1984b).

A recent National Research Council (1983) report concluded that VDT use is unlikely to be associated with increased risk of ocular disease or abnormalities; it noted that the radiation levels emitted by current VDTs are far below current U.S. occupational radiation exposure standards, are generally lower than the ambient radiation to which people are continually exposed, and are unlikely to be hazardous. The Research Council panel was less sure about issues of visual fatigue, muscular and skeletal discomfort, and stress, both as to the extent of these conditions in VDT-intensive jobs and as to the causal role that VDT use plays in their occurrence.

VDT users report higher rates of frequent eyestrain, muscle strain, tension, anger, and depression than nonusers and are more likely to have been treated by a doctor for a vision problem (9-to-5, 1984a). But, as the Research Council panel suggested, at least some of this effect is due to occupations: clerical workers, who are the most likely to be using VDTs, have the most symptoms. Occupational effects are substantially larger than VDT effects (9-to-5, 1984a). As emphasized above, moreover, in many cases VDTs are introduced into jobs that would be poor regardless of technology. In addition, some of the problems are due to the tasks people perform, not to the technology they use; tasks requiring close visual work that does not use VDTs produce similar symptoms of ocular discomfort, difficulty with vision, and temporary changes in visual function.

The physical environments in which VDTs are used also contribute to the problems workers have with them. Too often VDTs are used in bright rooms with glare that cuts contrast and makes screens hard to read. Tables may be the wrong height for reading and typing and seats may be uncomfortable for extended periods of sitting and too inflexible to fit the contours of a particular worker. A number of guidelines and standards now exist to give guidance to both equipment manufacturers and equipment customers about the design of workstations and workplaces for VDT-intensive jobs (e.g., Armbruster, 1983; Helander and Rupp, 1984; M. Smith, 1984).

This is not to say that the VDT itself is unimportant. As Stark in his dissent from the National Research Council report noted, "I have never seen a video display terminal that was nearly as legible as the ordinary pieces of typewritten paper or copied reports that circulate in our paper world" (National Research Council, 1983: 235). Working many hours per day on a hard-to-read and inflexible device may be sufficient to cause a number of physiological and psychological complaints.

Economic Considerations

Two issues dominate concerns about the impact of technology on economic aspects of employment quality: compensation and job security. To the extent that technology increases productivity, workers expect to share in the economic gain (see Murphree, 1985; 9-to-5, 1985). In the long term, increases in productivity, whether caused by characteristics of the worker (e.g., education) or by decisions made by the employer (e.g., use of technology), generally translate into increases in wages. But this longer-term equilibrium between productivity and wages may mask shorter-term perturbations during which workers feel they are not being compensated for the increased number of pages they type, customers they service, or analyses they perform.

Employees also expect to be compensated for specific skills they must acquire to use new technology. Training other workers is one especially important work responsibility that is associated with the introduction of information technology and that may be hidden and not directly compensated. For example, in the Kelly secretarial survey (Kelly Services, 1984), more than 50 percent of respondents reported that teaching others to use the word-processing equipment was a regular part of their jobs. On the other hand, employers may be using technology in order to reduce the skill requirements of jobs in an effort to reduce labor costs or may feel that technological literacy, like literacy in general, is a basic and noncompensable job requirement. The economic value of specific technological skills certainly interacts with their supply. Employers may need to pay a premium early in the life cycle of the technology, for example, when hiring word-processing operators in place of typists, independent of the skills needed to operate computers or typewriters. Such a wage differential is likely to decrease as more people with technological skills enter the labor market or more current employees learn new skills.

Employees also expect to be compensated for the general education levels that technology-intensive jobs may require (Noyelle, 1985). In the 1970s, however, the increase in wages per year of college education declined; while some observers expect improvement in earnings return in the 1980s, there is still some uncertainty about effects of education (Freeman, 1976).

Regardless of the impact of information technology on levels of employment and the structure of occupations in the economy as a whole, within specific firms and agencies technology has been used to reduce the absolute number of jobs and to redistribute employment among occupations. For example, Bikson (1986) found staffing reductions associated with new technology in about one-third of the offices she studied. When staffing changed, the numbers of clerical workers decreased and the numbers of technical workers increased. Certainly, as demonstrated in the previous chapter, the demand for some occupations is likely to vanish, even within particular firms. Employees' fears of job loss and employment dislocation that are associated with information technology can be reduced by a number of employer policies. The feasibility of these policies depends on business and other conditions. For example, employers can strive to ensure continued employment for employees whose jobs are eliminated or changed by technology by giving them preference for new openings and by providing or subsidizing training that is needed for new or changed jobs. Staff reductions, if they are needed, can be attained through attrition rather than layoff.

Conclusion

This section has illustrated the range of outcomes that can occur when computer and telecommunications technologies are introduced in women's jobs. The literature is rich with examples in which technology has been associated with either increased or decreased job quality. However, the lack of depth and uneven quality of the data prevent definitive conclusions. Overall, no compelling evidence was found that white-collar work in the aggregate has been getting either better or worse with the introduction of information technology. Because innovations can be implemented in broadly different ways, the major determinant of the effects of innovation appears to be management's preexisting employee policies. The examination of several specific areas of concern about information technology—changes in skills and job content, working conditions, and economic considerations—does suggest that regardless of the overall direction of change associated with technology, managerial workers generally fared better than clerical workers.

Although the predominant effect on employment quality of introducing information technology into the workplace cannot be determined with certainty, the range of outcomes found indicates that technology can be implemented in both job-enriching and job-degrading ways. Currently available hardware and software can be used with a great deal of flexibility. Identical hardware and software introduced into different organizational settings have different effects. The same equipment in two locations can be seen as reliable or unreliable depending on social setting (Blomberg, 1986). Identical equipment can be distrusted by workers or welcomed; it can be introduced with worker involvement or passivity depending on previous management styles and local traditions (Gurstein and Faulkner, 1985). Within a single job in a single industry, management practices and methods of work can be much more powerful influences on employees' job satisfaction than are the technological tools used to do the job (Herman et al., 1979). In sum, it is clear that any technology can be used in a number of ways and that social choices about its use are genuine.

IMPLEMENTING TECHNOLOGICAL CHANGE AND IMPROVING EMPLOYMENT QUALITY

This section first considers in more detail the process by which information technology is introduced to the workplace, focusing on the factors that lead technology to increase or decrease the quality of employment, especially for women clerical workers. Among the factors examined are the role of upper- and middle-level managers in decision making, including women's influence, and the economic, technological, and organizational constraints on managers' ability to consider and respond to employment quality. In brief, this examination, which is based on admittedly limited data, finds that the interests of managers and workers are likely to differ. The section then considers some mechanisms that can be used to ensure that workers' interests are represented in technological decisions.

THE ROLE OF MANAGERS

The Dominance of Management

Currently, upper- or middle-level managers are often primarily responsible for deciding to introduce information technology in white-collar work, for determining how much and what kind of technology to adopt, and for influencing many of the details of implementation. The limited evidence available suggests that the decision to introduce new technology is typically dominated by economic considerations and that implementation decisions focus on product selection. Both Johnson and her colleagues (Johnson et al., n.d.; Rice et al., 1983), studying organizational issues in 200 word-processing centers, and Bikson and her colleagues (Gutek et al., 1984; Bikson, 1986), studying the implementation of office automation in 55 offices, conclude that managers primarily initiate and influence the introduction of technology and that hardware and software considerations dominate the decisions. Issues such as employee attitudes, skills, and behaviors or organizational effects are rarely considered important.

According to the 9-to-5 survey on stress (1984a), almost two-thirds of the women who used automated equipment reported that they had no influence over the design, choice, or conditions of use of their equipment; female clerical workers reported substantially less influence than female managers. Clerical workers are more likely to be involved or at least consulted about equipment in small rather than large organizations, but they rarely serve on vendor-selection committees or make final decisions in any companies (Minolta Corporation, 1983).

Even when managers are concerned about the human elements of new technology, the scope of their concern seems circumscribed. For example, Westin et al. (1985) interviewed both end-users of technology and managers in 110 business, government, and nonprofit organizations that were reputed to be advanced and active users of office automation; most were selected because they had "good human resource policies" (Westin, 1985:4). In this highly selective sample, policies aimed at good ergonomics were widespread—better terminals, user-friendly software, adjustable workstations, and comfortable working environments—but even in these organizations, considerations of hardware and software dominated. Policies directed at job design—variety in tasks, discretion in pacing, or fair work standards—were less frequent and occurred mainly if human resource personnel joined data-processing and office automation personnel in planning for office automation. Almost half the sites that Westin and his colleagues visited had some formal employee participation programs (e.g., quality circles), although not necessarily oriented around technology. In only a quarter of the sites were such women's issues as sex-segregated work groups, career ladders in clerical work, or pay equity addressed directly.

The dominance of management in the introduction of technology is, however, not inevitable. The adoption and implementation of some technologies—e.g., bullet-proof vests by police officers, microcomputers in schools (Yin and White, 1984), and instruments in laboratories (Von Hippel, 1978)—have been initiated by workers. In Europe, as discussed below, users' interests are represented more directly.

Women's Influence

Compared with other workers potentially affected by information technologies, women have less influence over their use in the workplace because of their low position within office hierarchies, their relative lack of technical expertise, and their relative lack of professional and union representation. Women are less likely to be in senior management positions in organizations that implement technology and thus not in a position to influence its use. In 1984 women occupied only 34 percent of managerial positions in the United States but more than 75 percent of administrative support positions (Bureau of the Census, 1985). And, as noted above, clerical workers have much less influence in office technology decisions than do managerial and professional workers. Given the ubiquity of the sex segregation of jobs within firms (Bielby and Baron, 1985) and the small number of women in managerial positions, many of the managers who make decisions about the technology that women will use have never held a job like the one in which the technology is being introduced. Thus, they are likely to have difficulty identifying both the full range of tasks that the technology needs to support (Suchman and Wynn, 1984) and the full impact that the technology will have on the quality of employment.

Job segregation and less technical training make women less influential than men in technical decisions about hardware and software both in research and development firms and in the organizations adopting technology. While women are well represented in the computer professions, they tend to be concentrated as low-level programmers and documentation writers in insurance and banking, rather than as systems analysts and designers in research and development organizations or as managers in management information departments (Kraft, 1977; Strober and Arnold, 1985; Kraft and Dubnoff, 1986). Yet women clerical workers do have the detailed job knowledge that is necessary to design and apply computerized equipment in the most useful way (Suchman and Wynn, 1984).

Organized workers have more influence over conditions of their work than do similar nonorganized workers (Gorlin and Schein, 1984), and most women are not represented by labor and professional organizations. In general, except in the public sector, white-collar workers are much less likely to be represented by labor organizations than are blue-collar or service workers. Within the relatively unorganized white-collar labor force, women are somewhat less likely to be in unions than are men: in 1980, 14.7 percent of female white-collar workers and 16 percent of male white-collar workers were in labor organizations. In occupations that are rapidly adopting information technology, only 8.6 percent of female secretaries, typists, and stenographers, 4.5 percent of female bookkeepers, and 5.3 percent of women in retail sales were in labor organizations (Bureau of the Census, 1983:Table 729).

The Role of Enlightened Management

It is generally assumed that managers introduce technology in economically rational ways to ensure the survival and growth of their organizations: by producing new and better goods and services, by reducing costs, by increasing market share, by increasing profitability, and so on. In the optimal case, they manage in an "intelligently selfish" way that benefits themselves as managers as well as owners, customers, and workers. In this case, managers should strive to introduce technology in humane ways to maintain a stable, motivated, and effective work force. Managers can adopt values and procedures to introduce technology into the workplace in ways that use the technology effectively, aid the general welfare of the organization, and enhance employment quality (e.g., National Research Council, 1986). Methodologies such as the sociotechnical design of work systems (e.g., Mumford and Weir, 1979; Taylor, 1986) have been developed to combine the effective use of technology with its humane use.

There is much that an organization with a concern for employment quality can do to implement technology consistently with that concern. Westin (1985) and Bjorn-Anderson and Kjäergaard (1986), among others, have developed guidelines. The guidelines include building a concern for people into an overall office technology plan and obtaining top management's commitment to this concern; creating a task force to deal with people-oriented issues in implementation; formally involving workers in the design and implementation process; surveying ergonomic, health, and comfort conditions and upgrading offending situations, starting with the most technology-intensive jobs; establishing performance evaluations for technology users that are fair and avoid excessive monitoring; conducting employee-centered training for technology users and training their supervisors in methods to enhance employment quality; monitoring external developments in the policy, regulatory, and research communities to understand social concerns and to bring the organization into anticipatory compliance with the sound standards that emerge; designing technology-intensive jobs to allow variety, autonomy, and meaningful work; ensuring job security; and generally choosing and implementing the technology in a worker-oriented way.

For the use of VDTs, there is now sufficient consensus on standards for the functioning of the current generation of equipment to serve as a guide in selecting and deploying technology (Armbruster, 1983; Helander and Rupp, 1984; M. Smith, 1984). These standards encompass the issues of ambient lighting, glare, character size and legibility, character contrast, work placement, workstation flexibility, chair design, and the like. Undoubtedly, with changes in technological capabilities and with advances in research, the standards will need continuing updating.

Constraints on Managers

There are many reasons that managers do not act in the optimal way to maximize the effective and humane use of technology. Economic conditions, workplace cultures and traditions, technology, and conflicts between interest groups and values all influence the degree to which managers can and do emphasize employment quality in implementing technology.

Economic Conditions The economic condition of a firm is central to the technological implementation process and constrains the optimal case just described. Organizations that can pass any higher costs on to consumers—strong companies in a period of major industrial growth, companies with little international competition, and companies selling differentiable products and services that compete on design or quality—can often afford the resources necessary for implementing technology in humane ways. Conversely, weak companies—those in declining industries, those with major foreign competition, and companies selling commodities that compete primarily on the basis of price—may believe they cannot afford a longer-term view toward implementing technology humanely. The panel emphasizes, however, that it has found no evidence that implementing technology humanely is more costly and, indeed, some evidence that it is cost-effective (Commission of the European Communities, 1984).

Improved labor relations, decreased employee turnover, and more efficient operating systems are some economic benefits of taking quality of working life into account (Walton, 1975). As an example, Mirvis and Lawler (1977) have calculated the dollar savings (primarily from reduced training, less turnover, and lower absenteeism costs) of increasing average job satisfaction in one bank that they studied. It is likely, as Baran (1985), Bjorn-Anderson and Kjäergaard (1986), and Taylor (1986) argue, that a motivated and effective work force is especially important in an automated workplace because mistakes are more costly; errors have wider impact as shared data pass through an organization.

Organizational Culture and Behavior Of course the introduction of technology does not occur in an organizational vacuum. Even in his exemplary sample, Westin identified a minority of organizations in which top management's staffing approach was to encourage high turnover and low pay in the clerical labor force. These policies existed prior to the implementation of new technology and were the milieu into which technology was introduced. As Westin (1985:29) concludes:

[T]he advent of office systems technology was almost never the source of the poor practices. These managements had applied harsh personnel practices and engaged in sex discrimination before they had installed VDTs. Now they were extending their basic approaches into new-technology work settings [emphasis in original].

In addition, even in the context of good personnel policies, managers often have mixed motives and a lack of knowledge that leads them to decisions that fail to optimize their firm's welfare (see, e.g., Kling, 1980). For example, managers at different levels in an organization vie with each other for resources and influence on the direction of the whole organization and its components. Computerization is but one tool that managers can use in this competitive process.

Technological Constraints Except for specialized systems for relatively large organizations, the design and development of hardware is often controlled by equipment manufacturers and their research and development specialists, who are outside the workplace where the new equipment is to be used. Although software is more likely to be developed in-house by the data-processing departments of large corporations, much of the generic software (e.g., word processing, data-base management, graphics, spread-sheets, computer-aided design) used by the typical clerk, secretary, manager, and professional will have been developed by outside software vendors. Hence, the purchasers of technology have little direct control over the products available to them.

In general, the design of hardware and software is constrained both by the state of the art in technical fields and by commercial, social, and practical forces. For example, the larger screens, crisper characters, graphics capabilities, intelligent error handling, and user friendliness that have improved the use of computers in recent years have depended on a number of technological advances: the development of microcomputers, decreases in computer-memory prices, increased transmission speeds between video displays and central processing units, and refinements in software. The recent changes in clerical work in the insurance industries depend on hardware and software improvements that allow integration of formerly separate data bases (Baran, 1985).

Improvements in technology are not necessarily technologically determined, however. The most effective designs are likely to have been informed by interactions with potential users of the product (e.g., Maidique and Hayes, 1984; Maidique and Zirger, 1984). This interaction often occurs through usefulness and usability testing in developers' laboratories, through consultation with clients for whom a system is being developed, through the involvement of users in the development process, and through market feedback—some designs sell well and others do not. It is also possible for potential purchasers and users to influence design more directly. Large purchasers have enough purchasing power to set requirements on manufacturers for their automated equipment. Smaller purchasers and worker organizations can share evaluations of automated systems, as the National Education Association does for educational software and 9-to-5 does for VDTs.

In addition, some hardware and software have been designed to allow flexibility for the end-user. For example, some computer displays have adjustable character sizes and fonts as well as screens that rotate to reduce glare, and some computer programs adjust to the expertise of the user and allow the user to create new commands or to redefine old ones. Such flexibility is important to accommodate the personal preferences of individual users.

Constituent Conflicts and Value Contradictions   It is inevitable that the interests of parties involved in the introduction of technology into the workplace will differ. As stressed above, the decision of top managers to invest in technology often stems from a desire to reduce labor costs or to increase market share or product quality. Westin (1985) notes that, in the sites he visited, managers saw the balance between needs for cost control and needs for an effective work force as crucial. When vast sums are at stake, one can imagine that employment quality may take a back seat to more immediate economic concerns. For example, in 1983 the Bell operating companies employed approximately 35,000 directory or information assistance operators, who handled approximately 5 billion calls per year. In deciding to invest in new technology, the telephone companies use as a rule of thumb that a second of an operator's time costs 1.1 cents, which translates to $55 million per second for all directory assistance calls. These figures provided managers with powerful motivations to adopt technology and work procedures that reduce the time operators spend per call. The recent introduction of machine-generated reporting of telephone numbers to customers saves on average seven seconds per call; the dollar savings if this were introduced in all telephone companies would be more than one-third of a billion dollars per year. These potential savings, of course, must be weighed against increases in capital expenses and technical support; and, as Iacono and Kling (1986) note, technology has traditionally been touted for potential labor savings that are often not realized.

Even when the motivation for adopting technology is improved product or service quality rather than labor savings, effective use of technology may conflict with improved employment quality. For example, in the early 1980s the U.S. Social Security Administration changed the computer support for some of its claims representatives from a batch-oriented system, in which claims representatives entered a request and received a response approximately four hours later, to an on-line system, in which the response came within four minutes. With the on-line system, claims representatives could do a much better job of interviewing clients and establishing their entitlement to social security benefits in a timely way. But, as a result of the system, the claims representatives handled more cases per day, had more mental strain symptoms, and greater absenteeism (Turner, 1984). In this case it appears that better tools made worse jobs; increased interaction with clients and their problems and the difficulty of making decisions about them decreased job quality. Of course as the urban banking example of worker participation in implementation pointed out (Center for Career Research and Human Resources Management, 1985; reported in Chapter 2, this volume), the trade-off between improved productivity and improved employment quality is not inevitable. But even this generally favorable introduction of information technology was perceived as benefiting clerical staff more than it benefited the managerial staff, demonstrating the difficulty in making changes that have uniform benefits for everyone affected.

THE ROLE OF WORKERS

Although there is much that employers can do to implement technology in ways that preserve or enhance job quality, many factors constrain the ways in which computer and telecommunications technologies are introduced. In addition, as noted above, the goals that even the best-intentioned managers attempt to achieve are not necessarily compatible. Productivity, service quality, and employment quality do not necessarily go together. All of this implies that workers cannot simply rely on the goodwill of employers to ensure that technology is used humanely. Given the potential conflict between implementers of technology (usually managers) and workers, especially women workers, who use it on a day-to-day basis, workers need mechanisms to represent their interests in decisions affecting employment quality. This section first describes types of worker participation in the design and implementation of technologies, then considers their general effectiveness, and finally provides some detailed examples of participation by organized workers.

Worker Participation in Technology Design and Implementation

The long tradition in American management—from the early joint employer/ employee committees in the 1820s and 1830s (Guzda, 1984), to the Morse and Reimer (1956) study in the 1950s, to the present day (e.g., Katzell and Guzzo, 1983)—has documented that worker participation in what are traditionally thought to be managerial decisions (including work scheduling and the design of job activities) can lead to increases in workers' satisfaction with their jobs and to increases in organizational effectiveness. Worker participation affects employment quality and job satisfaction in two ways. First, it changes the contents of the decisions, because it provides a mechanism through which workers' interests are represented. Workers who will use information technology are often the only ones in a position to know enough about their jobs to design technology that articulates with those jobs; thus, their participation enhances the effectiveness of the new technology. Participation also helps ensure that information technology is not used in ways that decrease employment quality. Second, participation in decision making is intrinsically satisfying for many people and leads to increased commitment to decisions simply as a result of the process by which they were reached. Because worker participation generally has had positive effects in other contexts, it is likely to have positive effects when applied to technology as well.

Workers' participation can take two basic forms: informal participatory practices, which are the constantly evolving activities, knowledge, and expertise that workers bring to bear on many workplace policies; and formal rights giving workers an explicit role to play in company decision making concerning new technology (Howard and Schneider, 1985). The two forms are complementary. Without formal rights, informal participatory practices can be undermined or ignored when workers' interests collide with those of managers or technology designers. Without informal participatory practices, formal rights are rarely fully realized.

Many mechanisms exist for involving workers in design and implementation decisions. One of the most effective has been for workers to actively propose designs at both the research and development stages and then iteratively provide feedback for modifying the design of new generations of hardware and software products. This often happens in the professions, where workers have more control over the development of technology because of their status and expertise. As described in Chapter 2, such involvement is one route through which nurses have influenced the design of information technology in the medical professions. They communicated with each other about the technological possibilities, consulted with technology vendors, and became developers of technology themselves. It has also happened among craft workers where, for example, printers have joined with computer scientists to design an integrated text-and-image processing system for newspaper text entry, image enhancement, pagination, and layout (Howard, 1985).

It is instructive to consider the way in which computer programs that aid decision making about clients or patients have been used in different occupations and so have had different influences on the design and implementation process. For example, for physicians a program might accept as data a patient's symptoms and risk factors and the probability of a disease in the relevant population to categorize a patient as diseased or not. For a customer service representative in a utility company, the program might accept as data a customer's payment history with the company, the size of the current bill, and the probability of nonpayment in the general population to categorize the credit risk of a customer. While software for disease and credit-risk assessment could be very similar, they are in fact very different. Physicians' software is generally not used to make diagnosis but to train medical students in diagnosis (Richer, 1986), to provide them with graphics tools allowing them to see the implications of their tests (Cole, 1986), and to offer a system that provides a second opinion with explanations of discrepancies between a physician's judgment and that of the software (Langlotz and Shortliffe, 1983). In all of this software, the physician has access to the data and the rules that the software uses in ways that allow the physician to understand and challenge its assumptions and conclusions. In contrast, for a customer service representative in telephone companies (Dumais et al., 1986), the software schedules a service representative's telephone contacts with customers based on such data as the size of the current bill, the length of time that a bill has been overdue, and the action to be taken—without providing the representative fast access to either the data, the scheduling rules, or the application of the rules to a particular customer. Appelbaum (1984) and Baran (1985) find a similar lack of access to the software's decision-making rules in the insurance industry: the software makes decisions on whether to insure and at what rate based on data the insurance clerk enters. The clerk has very limited ability to interact with the software in understanding how decisions are made.

There are many reasons for the differences in these systems, but one is that physicians have been more involved in the design of the software as designers, consultants, and research funders. In addition, since physicians are generally discretionary users of computer software, they can limit their use to systems that aid them without usurping what they consider to be their responsibilities, autonomy, and expertise. Thus, to gain physicians as clients, software developers provided systems that do not impinge on physicians' autonomy.

Other mechanisms for increasing workers' influence include legislation and regulation, for example, codetermination legislation that requires employers to inform workers about plans for future developments and to initiate discussions and negotiations on new technology before the changes or final decisions take place (e.g., 1977 Swedish Act of Codetermination); negotiated collective agreements about technology on national, corporate, or establishment levels (e.g., the 1984 typographers' contract discussed in Chapter 2); and more informal discussion and influence, both in the context of worker representation on implementation committees (e.g., Yin and Moore, 1984) and the normal informal consultation between managers and other workers that is common in small offices (Evans, 1983).

The Effectiveness of Worker Participation

Evaluations of the effectiveness of user involvement in the design of information technology are sparse. They suggest that users' involvement in the design and implementation of information technology can be a mechanism to improve its effectiveness and to improve the jobs in which it is used. For example, in her study of the design and implementation of computer systems in six British organizations, Mumford (1981) notes that effort is expended on human goals such as job satisfaction during the system design and implementation processes only if they are explicit design goals. Those who participate directly in the early stages of the system design process are able to influence the nature of the goals set and hence the way in which the system ultimately operates. If users participate in or control the design of the system, the goals that are important to users, such as job satisfaction, are more likely to be attained. In several of the organizations that she studied, however, managers and system designers, not users, were responsible for trying to design information systems that would meet users' needs.

The effectiveness of worker involvement in information system design and implementation depends vitally on the organizational context in which participation takes place and on the procedures through which it is accomplished. Mechanisms for worker participation vary in the scope of workers' interests that are represented and the degree of influence that workers have. The Commission of the European Communities (1984) reports on the concept of "design space" developed by J. Bessant to help understand the range of issues open for negotiation when implementing technology; the concept has been elaborated by a group of researchers investigating worker involvement in five European countries (Commission of the European Communities, 1984). Design space can be thought of as the range of choices about technology, work, and work organization that are possible. Constraints that limit the choices include the available technology, the regulation of industrial relations, company characteristics such as industrial sector, size, type of production process, and such environmental determinants as economic climate, competitors' behavior, state of the market, and availability of resources.

It is important to note that the range of available choices is itself a point for negotiation. The range of issues that can be included in the design space is broad. Analysis of European cases shows that it includes at least the following (Commission of the European Communities, 1984): changes in the organization of the firm and its investment in automation (e.g., company organization, degree and timing of automation), changes in the organization of work (e.g., division of labor, job composition and skill requirements, job autonomy and cooperation, pace of work, training, recruitment and promotion, grading and pay), and the actual hardware and software technology.

Decisions made early in the implementation process constrain the choices that can be made later (Mumford, 1981). For example, a decision to hire specialists to perform data-base searches in a law office constrains the work organization of legal secretaries and, therefore, the nature of the technology they use. This funnel of choice implies that if worker involvement in technological change is to be effective, it must include involvement at early stages.

Because users of computerized information systems are sources of expert information about the jobs that are to be automated, political actors whose acceptance of or resistance to information systems can determine their success, and workers whose knowledge, training, and skill in actually operating a system will determine if it is used effectively, a number of developers of information technology have involved potential users of their systems in design, implementation, and training. They have solicited their opinions during the design of the systems, informed and involved them during the implementation process, and provided them adequate training after the systems have been installed. The Digital Equipment Corporation, for example, advocates a user participation methodology developed with Enid Mumford and known as ETHICS—Effective Technical and Human Implementation of Computer-Based Systems (Mumford and Weir, 1979; Bancroft, 1982; Bancroft et al., n.d.). Once an office is slated for automation, a broad-based design group of workers analyzes their own workplace, proposes alternative ways of organizing work tasks, and helps select appropriate technology. This design group works in consultation with technical experts and under the guidance of a management steering committee that sets the design group's charter. But as Howard and Schneider (1985) note, these approaches are often ineffective in representing workers' interests because they are often used as marketing tools by vendors to sell products to skeptical buyers, rather than as genuine mechanisms to solicit information from users. Second, the "users" who are consulted are often first-line managers, that is, users' supervisors, rather than the day-to-day users of the technology. Third, the scope of what user groups are allowed to consider is constrained by management oversight, and many of the crucial decisions about need for technology, spending levels, personnel requirements, layoffs, and job design are made without users' input or influence. Fourth, user groups that operate at the level of an establishment or even a firm have limited scope for influence and can rarely enlarge the set of technological alternatives available in the marketplace from which they must select. Finally, because such user groups operate at the convenience of management, they have no formal rights in corporate decision making to rely on if the informal consultation arrangements founder.

Organized Worker Involvement

A number of techniques have been used in the United States to generally increase workers' involvement in managerial decision making. These include, among others, problem-solving groups and quality circles, whose purpose is to identify and analyze workplace problems and present solutions and implementation plans to management for approval; autonomous work groups, which collaborate in the completion of work tasks and have responsibility for implementing solutions and controlling the day-to-day scheduling, standards, and flow of their own work; and business teams, in which workers participate with managers in decisions affecting product development (Gorlin and Schein, 1984). Worker participation programs have been used primarily in manufacturing companies, generally with union representation. They have rarely dealt with technology design, often focusing on local quality-of-work-life issues, including employee morale, safety and environmental health, scheduling arrangements, absenteeism, overtime, bonus payments, job alignment, and, occasionally, product quality.

Both in the United States and in Europe, unionized workers have negotiated technology agreements that provide mechanisms for union involvement in the implementation of technology in their industries (see Chapter 2). Both procedural provisions about the manner in which technological change should occur and substantive provisions about the content of technological change are often included in these agreements. Typical agreements include the following (Evans, 1983):

- Management provides workers advance information about plans for technological change.
- Joint management/union committees discuss, monitor, and negotiate change at the corporate and establishment level.
- Unions have access to outside expertise to guide their technological decisions.
- Employers provide employment security following the introduction of new technology.

THE QUALITY OF EMPLOYMENT

- Employers provide adequate retraining to workers whose jobs are changed or eliminated.
- Joint management/union committees monitor the impact of new technology on the quality of working life.

For example, in the United States the 1980 contract between the Communications Workers of America (CWA) and AT&T set up shop-floor committees on the quality of work life, which could address local technology issues, and higher-level technology change committees, in which, according to the language of the formal letter of agreement, the unions could "discuss major technological changes with management before they were introduced" (reported in Howard and Schneider, 1985). The quality circles in the Bell system, like many others, have dealt primarily with matters affecting the immediate workplace environment, such as scheduling, management attitudes, and office environment, and, according to Howard and Schneider (1985:14), "have been an important vehicle for building workers' informal participatory practice at the local level." Technological change, however, has generally not been addressed.

The technology change committees, intended to be a forum for discussing broad policy issues about technology, have been generally ineffective: a 1983 study (reported in Howard and Schneider, 1985) found that during the first three years of the program nearly two-thirds of the committees had yet to meet. The agreement by AT&T to provide the unions with six months' advance notice of all major technological changes provided insufficient time for union influence on technologies that often required years between planning and implementation. In addition, the notification agreement gave unions no formal rights to participate in the conception, design, or testing of new technological systems. Finally, neither the union nor corporate representatives on the committees had timely information about the types of systems that were being developed. The corporate representatives were typically labor relations personnel with expertise in dealing with unions, not technology. Engineers, systems designers, and technical managers were rarely committee members.

European, especially Scandinavian, countries have higher levels of unionization than the United States; relatively homogeneous societies (compared with the United States); a tradition of industrial democracy, with a 20-year history of worker participation in many aspects of workplace decision making; and the reinforcement of contract provisions by parallel provisions in laws and regulations. As a result, technology agreements in Europe have been more elaborated and much more effective than in the United States. In contrast to the relatively limited and unsuccessful U.S. experiences in worker participation in technological change, many European examples show that worker participation can have substantial impact on how technology and work is designed.

Much of the evidence about participation in the design of technology comes from the Scandinavian factory experiments of the 1970s (Norstedt and Aguren, 1973; Aguren et al., 1976; Thorsrud et al., 1976). In these projects, factory workers, both directly and through union representation, participated along with management and staff groups in the design of the workplace, the design and selection of the technology used to do jobs, and the design of jobs. They also made many of the day-to-day decisions about how tasks are organized and work is scheduled. For example, in the Volvo factory at Kalmar, Sweden, social goals were identified early (Pehr Gyllenhammar, quoted in Aguren et al., 1976:5):

to organize automobile production in such a way that employees can find meaning and satisfaction in their work . . . [to] give employees the opportunity to work in groups, to communicate freely, to shift among work assignments, to be conscious of responsibility for quality, and to influence their own work environment.

These goals led groups of foremen, technicians, architects, and trade unionists to emphasize team assembly when they designed conveyance and assembly equipment, its controlling software, and work organizations. Within the constraints of the installed equipment, the teams could decide how assembly would be done. Evaluations suggest that by conventional criteria of quantity, quality, and costs, the new methods are acceptable, equaling conventional assembly methods. Some of the traditional problems of assembly work—absence and turnover—are reduced. In addition, the new methods changed the content of work, giving workers longer work cycles, job rotation, and more control over task allocation, for example. As a result, workers were very satisfied.

The involvement of the Norwegian Bank Employees Union (BNF) in the banking industry provides an example more relevant to women and information technology (Howard and Schneider, 1985). Women make up about one-half of the union membership. They generally hold the lowest-paying jobs, 68 percent have no professional training, and nearly 70 percent work part time. To ensure that these women would not be trapped in low-wage, dead-end jobs as technology transformed the banking industry, the union emphasized access to training in its technology policy. Union and management set aside 40 percent of the places at their joint industry training center for women, and the union negotiated special arrangements for training working mothers.

The BNF developed elaborate structures to guarantee workers' participation in technological development, and union representatives participated actively in the design of new computer-based systems. For example, in 1982, 80 bank employees, working in 10-person groups, developed the preliminary software specifications for a $70 million computer system for savings banks. When system designers proposed a new automated loan processing system, a union-management team evaluated its consistency with organization goals. They suggested such changes as retaining decision-making power to grant or withhold loans with loan officers rather than delegating these decisions to a computer algorithm. In addition, union representatives maintained informal relationships with bank technical personnel, both to educate them about the union's social goals and to influence the design process. Similar strategies are reported for female-dominated workplaces in Norway, such as the postal services and public libraries (Bermann, 1985).

These examples show that union involvement can be a mechanism for worker participation in technological design and implementation. However, given the large differences between European and U.S. societies, one might be skeptical about the applicability of the European experience to our own (but see Mumford and Weir, 1979, and Mumford, 1983, for a counterargument). Different mechanisms for worker involvement may be necessary in the United States, especially to ensure worker involvement among the highly nonunionized female labor force in service industries.

Conclusion

In the United States managers are generally responsible for decisions to introduce information technology. While there is much that enlightened managers can do to ensure that information technology is used in ways that enhance jobs, most managers do not include users of the technology in the decision-making process and rarely treat employment quality as an explicit and important determinant of their decisions. Women clerical workers in particular have little direct influence in implementation decisions, because of their position at the lower end of the organizational hierarchy, their lack of collective representation, and their generally lesser technical knowledge. The inevitable differences in interests between managers and workers suggest that workers need mechanisms to participate in technological design and implementation. Worker participation programs in both the United States and Europe have been used to improve both general organizational effectiveness and employment quality.

Worker involvement in the design and implementation of technology in general, and union involvement in particular, are not panaceas, however. Unions can be as autocratic and unrepresentative of their members as managers are of workers (Lipset et al., 1956). Direct participation by individual workers in the technological decisions that will affect their own jobs implies a limited scope for their influence, limited to the establishment in which they work and to technology already developed. If participation is to be effective, it must be involving, not merely formal, and strong ties between representatives and their constituencies must be maintained (Katz and Kahn, 1978). Managers, with their superior power and resources, can often manipulate workers while providing a facade of participation. These asymmetries in power may be especially important during a period of rapid technological change.

However it is done, true participation is very difficult to achieve. As Katz and Kahn (1978:767) conclude:

Research, experience, and experimentation agree that participation in organizations can be increased substantially over conventional bureaucratic practice, with positive effects. The data also agree that the process of attaining such increases is not easy, that it implies changes beyond what most organization leaders are prepared to initiate spontaneously, and that some conventional positions of supervision and technical specialization are explicitly threatened by the participative process.

Whatever mechanisms for participation are developed, they must, in the long run, contribute to economic viability. Worker participation, management interests, employment quality, productivity improvement, and competitiveness are all factors that enter into the long term viability of economic activity.

5

Conclusions and Recommendations

SUMMARY

This report has examined the effects of technological change on the quantity of women's paid employment opportunities and on the quality of their jobs. Because so many women work in clerical occupations, it has focused primarily on innovations in telecommunications and information processing and on their impact on clerical work. New microprocessing and telecommunications technologies have been introduced and applied rapidly in offices in the last 10 years. This rapid change is part both of continuing technological change in the industries that employ clerical workers and of concomitant changes in work organization, which occur in response to new market conditions and other economic or political factors.

Much clerical work, especially in large workplaces but to some extent in nearly all offices, has been reorganized; the content of clerical jobs has been modified; the geographic distribution of jobs has shifted; and new products and services have been introduced, while others have disappeared or declined in importance. Some occupations have almost disappeared from the labor market and new ones have emerged. Overall, the total number of jobs held by clerical workers has not declined, but the rate of clerical employment growth has slowed relative to the growth of total employment. There have been both increases and declines in employment linked to the introduction of new technologies in a number of specific occupations.

The effects of technological change are difficult to identify and even more difficult to quantify for several reasons. First, technological changes occur within the context of many other changes—in economic conditions, international competition, and labor supply. Second, the data available for studying the employment effects of technological change are inadequate. Third, the changes of interest are still very much in process, and past changes set limits on the present and future. Fourth, the future economic environment in which the adoption of technologies will take place is uncertain. With these limitations clearly in mind, the panel has thoroughly considered the evidence available and offers its best judgment, as of 1986, as to what the next 10 years are likely to hold for women's employment.

With regard to levels of employment, the panel expects that current patterns of change will continue over the next 10 years. These patterns—the slowdown of growth in clerical employment, the on-going shifts in the relative importance of various clerical occupations, and the reorganization of work—together are of a magnitude that some workers will be seriously affected. Some may become unemployed. Others may experience skill mismatch between their capabilities and available jobs. New entrants may have difficulty finding entry-level jobs. However, most of these effects are likely to be transitional. In this chapter the panel recommends courses of action to assist workers, employers, government, and private organizations to respond to the changes identified in this report. Since there is some uncertainty about our expectation of continued moderate change, however, the panel also offers policy suggestions that should come into play if the future employment situation develops in a way that more severely affects clerical workers than is now expected.

The new technologies have the potential to alter job quality as well as job quantity. The panel has not been able to determine whether, on balance, the technologies have thus far decreased the quality of clerical work by increased routinization, fragmentation, or abusive electronic monitoring, for example, or improved it by facilitating intellectual challenge and growth of competence. Effects are inevitably mixed. The new technologies have been implemented in a wide variety of ways and will continue to be introduced in new and diverse situations. Several surveys show that workers are generally satisfied, and often happy, with the new capabilities because they have eliminated repetitive tasks, such as retyping manuscripts. But several surveys of health effects show that workers who use automated office equipment and report that they have strict production quotas or are monitored closely are more likely to indicate stress-related psychological and physiological symptoms. The panel's recommendations are aimed at minimizing the negative possibilities and improving job quality. They are primarily concerned with the transition period in which different changes may bring about temporary job loss for some workers and with particularly vulnerable groups. They also address possible reductions in job quality when technology is introduced to some aspects of clerical jobs.

Because of women's preponderance in the clerical work force and because of the importance of clerical work as an occupation for women, the trends in job

CONCLUSIONS AND RECOMMENDATIONS

quantity and job quality can pose special problems for women. Women are more likely than men to have received an inadequate education in mathematics, science, and technology. Hence, they may be less able to take advantage of new opportunities that involve technical skills. Women also still bear most of the responsibility for dependent care and household tasks. Thus, women facing unemployment, the need for retraining, or possible relocation may face additional barriers to successful negotiation of transition to new jobs. For various reasons—including the types of jobs they hold, opportunities for advancement, and their generally lower wages—women may not have received valuable training or education programs while employed and so may face the future with less training, fewer skills, and more limited resources than other displaced workers. Women may also have less experience in participating in workplace decisions that affect their job content and job design.

Minority women's problems and needs are even more acute because of their greater disadvantages in education and because of their place in the occupational structure. Their jobs, including those within the clerical sector, are clustered in relatively few occupations and industries, which in some cases are expected to be especially negatively affected by technological change. Physically handicapped or older women and new entrants may also experience disproportionate effects.

Although the panel expects that in some ways the new technologies will affect women and men differently, in other ways the effects on women workers occur primarily because of their roles as workers, not because they are women. Both women and men may experience a poor fit between skills developed in previous jobs and those in newly available ones. Women and men with adequate education and training in school or on the job will be able to adapt to the anticipated changes in the distribution of available jobs; they will be able to shift among clerical jobs or to pursue other jobs. Broadly speaking, technological change requires flexible responses from all workers. All workers can benefit from good basic education and additional education or training in science and technology. Therefore, the panel's recommendations address the needs of women as workers (and, hence, apply to male workers as well) in addition to women workers' special needs that stem from their present disadvantaged position.

Overall, most consumers, employers, and workers benefit from technological change through lower prices, new products, higher profits, and higher wages based on productivity increases; some workers and firms, however, are losers in the process. The panel has based its recommendations for private and public action on the long-standing tradition that when costs are borne disproportionately by a few, they are socially shared. There are several reasons for this tradition. First, whenever benefits are shared by many but the costs are borne by few, market mechanisms to distribute costs are unlikely to develop on their own. Second, there are economies of scale both in dealing with uncertainty and in collecting and disseminating information about societywide change. These reasons imply a role for government in monitoring the distribution of costs and facilitating sharing, which have been and continue to be functions of government at all levels. In addition, governments are major suppliers of services and major employers, and they are also large users of new technologies. Thus, they have the potential to act as a model of good implementation practices. Because of the preponderant role of the federal government and the national scope of many employers and unions, the federal government should be a key participant. The social sharing envisioned here has both a private and a public component. In addition to government monitoring, supervision, and support, collaborative efforts by designers and manufacturers of equipment, employers, educators, workers, unions, and women's organizations are needed to manage the changing number and distribution of clerical jobs and to alleviate possible reduced quality in these jobs.

The recommendations that follow address the two types of change discussed in this report—the quantity and quality of employment. They represent the considered view of the panel about the best ways to deal with the changes currently under way and expected in the next decade. The recommendations are organized by eight subjects: education, training, and retraining; employment security and flexibility; expansion of women's job opportunities; adaptive job transitions; identification and dissemination of good technological design and practice; worker participation; monitoring health concerns; and data and research needs. Each of the sections begins with a statement of the panel's major conclusions.

EDUCATION, TRAINING, AND RETRAINING

Technological change today is science-based. Technological change will continue to affect the quantity and quality of jobs in many ways. Substantial occupational shifts can be expected. A good preparation in basic education and particularly in science and mathematics is therefore critical. Education, training, and retraining both in schools and on the job should be viewed as life-long activities, because changing employment conditions will increasingly require flexibility on the part of workers.

The current rate of technological change and its diffusion throughout the economy is difficult to evaluate, but there is little evidence that it is faster today than in the recent past. The nature of technological change has been transformed, however; more than ever before, it is the product of systematically applied scientific knowledge. Advances in scientific knowledge continually generate the potential for new technical applications.

As in the past, technological change will affect employment quality and quantity; some jobs will disappear and others will appear; still others will be fundamentally altered. All workers should be prepared to adjust to changes in the job market. A strong, balanced, basic education is fundamental both to an individual's ability to adapt and to the economy's need for flexibility. The panel reaffirms the findings of the Panel on Secondary School Education for the Changing Workplace (Committee on Science, Engineering, and Public Policy, 1984) calling for school and parent efforts to develop sound work habits and attitudes; to prepare young women and men for participation in workplace decision making; and to provide an education that includes such skills as command of the English language, reasoning and problem solving, comprehension and interpretation of written materials, accurate and clear written and oral communication, computation, and basic understanding of scientific and technological principles and of social and economic phenomena. In addition, the panel believes that compensatory programs for disadvantaged students in such subjects as English, mathematics, and computer skills are desirable.

Women have often been disadvantaged in the skills required to compete successfully in the workplace. Schools, employers, and government programs have all contributed to this outcome; they have not always made programs equally available to both sexes. Furthermore, minority women have often not had adequate access to high-quality basic education, including scientific and technical training in school.

Some women who have had access to solid educational preparation have been discouraged or barred from benefiting fully from it. Girls and their parents have often not been aware of the consequences of their educational selections. Females have usually taken fewer mathematics, science, and computing courses in high school than males; have entered less technically oriented vocational programs; and have been less likely to complete college-level programs in scientific and technical fields. The differences in available opportunities and the presence of structural barriers to taking advantage of opportunities have far-reaching effects for women's work lives.

Once employed, women have not always had equal access to on-the-job training programs and to other training opportunities provided by employers. Employment-related training for women both provides an important resource to workers faced with job changes and contributes to employers' goals of building an experienced, flexible work force. Women's long-term attachment to the labor force has increased in the past several decades and, in similar jobs, women's turnover rates are equal to men's: consequently, investment in training for women is as likely to pay off as it is for men. Because of women's customarily greater dependent care duties, however, they may need additional help to manage dependent care when education or training takes place outside normal working hours or locations. (Men who are responsible for dependent care, of course, have the same needs.)

The panel recommends that:

- schools provide more equal opportunities to women students and create conditions that actively encourage them to take more mathematics, science, and technical courses;
- employers support (for example, through tuition reimbursement) the provision of both general basic education and specialized programs that are not necessarily immediately job related;
- employers make on-the-job training programs equally available to all workers, with appropriate help for those with dependent care responsibilities;
- employers provide special encouragement to women workers to participate in technical education and other training that will expand their knowledge and skills;
- government-sponsored technical training programs also particularly encourage women to take part and offer necessary support services for those with dependent care responsibilities;
- governments (especially at the state and local levels) and private organizations collect and evaluate data on the availability and quality of training and retraining programs with special reference to women. They should also fund demonstration projects examining the usefulness of such programs, disseminate information about successful models, and, when appropriate, provide continued funding for high-quality training and retraining programs.

EMPLOYMENT SECURITY AND FLEXIBILITY

Technological change brings about shifts in the demand for workers in different occupations and changes in the content of their jobs. Hence, it is important, both to a business organization's economic success and to workers' employment security, that both employers and workers be flexible in responding to change.

Recent technological change has already made some occupations, such as tabulating machine operator, mimeographer, and keypunch operator, obsolete; employment in others, such as bookkeeping and stenography, has declined substantially; and growth in some, such as bank tellers and data-entry clerks, has slowed. This pattern is expected to continue. Additional, as yet unidentified, jobs will disappear or be fundamentally altered, and new ones will be created. Some occupational effects may displace particular workers from particular jobs. Some employers, particularly large ones, may have several alternatives when faced with declining jobs in some areas and occupations, depending on their market position, their economic and human resources, and the qualifications and flexibility of their workers: they will be able to restructure jobs and reallocate work to preserve employment. Other employers, because of

CONCLUSIONS AND RECOMMENDATIONS 173

disadvantaged positions on the same variables, may be more constrained. The panel recognizes differences among employers that affect their ability to implement its recommendations. It nevertheless encourages both employers and unions to take a more active role in anticipating technological change and in developing approaches that minimize employment insecurity. When employment security cannot be maintained, sufficient advance notice to workers is critical.

The panel recommends that:

- employers seek to maintain continued employment of individual workers in the face of change, even if it is not possible to guarantee continuity in their specific jobs;
- employers monitor the application of new technologies, the resulting potential job changes, and respond, to the extent possible, with policies (for example, job rotation, retraining, and assistance with geographic relocation) that provide secure employment for individual workers;
- unions and other worker organizations encourage worker flexibility in return for employment security and explore new membership forms that do not hinder such flexibility (for example, seniority benefits that are not negatively affected by the loss of a specific job, job change, or even employer change).

EXPANSION OF WOMEN'S JOB OPPORTUNITIES

Although technological change is not likely to cause large declines in clerical employment, growth will slow and occupational shifts among clerical specializations are likely to be significant. Women workers will be more affected by these changes because they work disproportionately in clerical jobs. Moreover, because clerical employment has provided jobs for many women entrants and reentrants over the past several decades, some new women entrants and reentrants are likely to have to seek jobs elsewhere. Thus, the effect of technological change in the clerical occupations on women's general employment opportunities will depend on their opportunities elsewhere. In particular, it will depend largely on the evolution of equal employment opportunity enforcement, economic growth, and women's qualifications.

The projection provided in Chapter 3 of what the panel judges the largest plausible negative impact of information technology on clerical employment concludes that by 1995 clerical employment will have lost at most 2 percentage points of its share of total employment: 2.0 million clerical jobs, or an increase of 10.5 percent, will have been generated relative to a growth in total employment of 20.4 million, or 22.2 percent. Compared with the actual growth of clerical employment between 1972 and 1982 of 4.1 million jobs, or a 29 percent increase, the "most plausible worst case" projection suggests an annual growth rate one-third as large, 0.7 percent per year for the 1982–1995 period compared with a 2.3 percent per year growth rate for the 1972–1982 period. The growth rate in clerical jobs, in this projection, would also be about one-half the growth rate of total employment. While in the past women's concentration in clerical jobs has worked to their advantage in terms of employment growth, in the future it can be expected to work to their disadvantage.

On the supply side, the rate of increase in the supply of women workers to the labor market will also slow. The number of women in the civilian labor force increased 36 percent between 1975 and 1985; the anticipated increase between 1985 and 1995 is 20 percent. Thus, while the rate of growth of clerical jobs will decline by two-thirds, the rate of growth of the female labor force is expected to decline by less than one-half. A potential oversupply is suggested. The supply of women to clerical jobs may be further reduced, however, because women can be expected to continue to train for and enter nontraditional jobs, as they have during the past 20 years and especially in the past decade. If unemployment does result, however, it is not likely to be solved by an exodus of women from the labor force. Because of their increased attachment to the labor force, women, like men, will continue to seek employment. On balance, considering both supply and demand, whatever unemployment problem might result from technological change in clerical occupations should be small.

The slowdown in growth of clerical jobs is likely to be smaller than often popularly imagined, for several reasons. First, the machines that capture the imagination take time to diffuse widely. Second, those machines frequently are not capable of replacing human skills to the extent their designers and adopters imagine or hope. Although some types of new clerical technologies (word processors and intelligent cash registers, for example) have diffused fairly rapidly, especially in large businesses, others (such as the fully automated workstation) are unlikely to become realities for more than a handful of workers over the next 10 years. Second, despite the growing availability of alternative technologies, the keyboard is likely to remain the dominant means of data entry. Third, networking between systems will remain a problem that requires both technical and political regulatory solutions. Fourth, the productivity gains that are possible from improvements in telecommunications and information processing will take many forms, in particular the development of new products and services and the improved quality of existing products and services; labor cost cutting is not likely to be the dominant response by employers across the economy.

The employment consequences of slower job growth in clerical work also will depend on the extent to which economic growth generates opportunities in other sectors, the extent to which desegregation generates opportunities in jobs that have been male dominated, and the extent to which women have the appropriate qualifications to benefit from these new opportunities. Although, on balance, the panel does not expect widespread unemployment problems, it is concerned that as women's financial responsibilities for themselves and their families have increased over time, so too may the consequences of their unemployment be more serious.

The panel recommends that:

- government continue to enforce equal employment opportunity legislation and related policies that enable women to find nontraditional jobs and that it continue to promote gender-equitable patterns in employment, in job training programs, and in technical, vocational, and general education programs;
- employers, unions, and educators actively seek to ensure opportunities for women in both nontraditional and new occupations, and, to this end, develop training programs and educational opportunities that enhance women's transfer and promotion possibilities.

ADAPTIVE JOB TRANSITIONS

Technological change can contribute to unemployment in several ways. Although our best guess is that massive unemployment problems resulting from technological change will not occur over the next 10 years, some job loss in specific categories, leading to some transitional unemployment, will occur. If economic growth is sluggish, the labor-displacing capabilities of technological change may contribute to a more serious unemployment problem. Workers are likely to need assistance in either case.

Technological change inevitably involves some skill mismatch as some occupations grow or emerge and others shrink or disappear. In the transition to the new types of jobs, there will be pockets of joblessness. Whether the unemployment problem is transitional or becomes more severe, shifts in the demand for labor could decrease the need for labor in several ways. Workers may be laid off from their current jobs; if voluntary turnover is sufficient, no layoffs need occur, but new workers will not be hired. New entrants and reentrants to the labor market may experience difficulty in finding jobs appropriate to their qualifications. And workers who voluntarily leave their jobs may experience difficulty in getting rehired.

Transitional problems require solutions in their own right, even if the magnitude of the overall unemployment problem is, as expected, not large. In most cases unemployed workers will have no continuing relationship with a previous employer. Some laid-off workers will not be able to anticipate going back to a previous job or to a new job with a former employer. Many of the unemployed will be new entrants or reentrants. Because such workers are bearing the cost of a social process (technological change and productivity improvement) that is potentially beneficial to all, public support for them is warranted. Women with limited educational backgrounds and those located in disproportionately affected geographic areas are especially likely to require support. Affected male workers will also need assistance.

Adaptive job transition programs for technologically unemployed workers should emphasize retraining or relocation support or both. An important component of such programs would be job search and placement aid. In modifying existing programs to meet the needs of technologically unemployed workers, or in developing new ones, there is a need for awareness of several issues that particularly affect women. Additional financial aid may be required for retraining and transition when the only household wage earner is a female with limited financial resources and substantial family responsibilities; dependent care is likely to be an important need. Women's geographic mobility may be constrained by the potential loss of location-specific support systems; relocation services should include assistance with finding dependent care. In the case of households in which both husband and wife have been employed, geographic mobility may be limited by dual-earner considerations; job search assistance should be provided for both partners even if only one was affected by technological change, in order to facilitate the adjustment of the household. And in the job search process itself, women, especially minority women, may need assistance in overcoming discriminatory treatment. They may not require retraining at all, but simply access to the full range of available jobs for which they have the required skills.

The extent of the employment impact of technology is dependent on overall economic conditions and public policy. Obviously, employment losses due to technological change are easier to cope with if the economy is growing strongly. The cost of change, in the form of public programs or private actions, is also better managed when the economy is growing. The panel made no attempt to predict the economic path of the United States over the next 10 years, but its judgments about the future are implicitly based on the assumption of a modest and steady increase in aggregate output and demand. If economic growth falls off steeply, pressure for labor cost cutting may become a more significant factor in determining how new technologies are deployed. An enormous reservoir of unexploited productivity gains exists with the new technologies. If these potential productivity gains are all deployed toward cost cutting, rather than on improving quality or introducing new products or services, the effect on employment could be severe. There are some indications that widespread cost cutting could occur, but, on balance, the panel believes it unlikely. Substantial unemployment could also come about for a variety of other reasons unrelated to technological change; these were not examined by the panel.

The recommendations that follow are designed to deal with the job transition problems that seem most likely, but they also provide models and information that would be useful if more widespread technological unemployment than is now anticipated occurs.

The panel recommends that:

- current state and federal employment programs improve their ability to deal with technological unemployment and develop new programs that pay special attention to the needs of women workers and new entrants, since they are likely to be disproportionately affected;
- local and state governments, employers, employer groups, women's associations, unions, and workers collaboratively plan and implement, perhaps through the Job Training Partnership Act and the Private Industry Councils, new job transition programs for technologically unemployed workers that take account of the special problems of women and families.
- the Women's Bureau of the U.S. Department of Labor, or other appropriate office, inventory and evaluate demonstration programs that are especially effective in dealing with the unemployment of women workers that occurs because of technological change; such an inventory would contribute to the management of a more severe unemployment problem, should it occur.

IDENTIFICATION AND DISSEMINATION OF GOOD TECHNOLOGICAL DESIGN AND PRACTICE

For all organizations, there are better and worse practices with regard to the design, implementation, and application of technology. No single set of practices is likely to be good for all organizations or for a single organization at all times. Nevertheless, the panel believes that identification of good user practices with respect to information technology would contribute to the formation of higher standards overall. By "technologies" we mean both new machines (for example, a word processor) and innovations in work organization (for example, remote transcription of dictation).

Alternative implementation approaches need to be identified and assessed for their effects on productivity, job quality, worker satisfaction, and so on. The design of equipment, in particular its ergonomic features (those that facilitate the use of the equipment by humans), also needs evaluation.

The panel recommends that:

- public and private organizations systematically assess current practices in the introduction of technologies and broadly disseminate their findings;
- employers, manufacturers, and designers, and their representative organizations, as well as women's organizations, unions, and educational institutions, take an active role in identifying applications that work better than others and disseminating information about them;
- designers and manufacturers of equipment and employers and workers who use it share their knowledge and experiences so as to facilitate the creation of sound ergonomic standards and practices;

- that governments, as major generators and users of equipment and software, facilitate information sharing and act as models to other institutions through their own adoption of good practices.

WORKER PARTICIPATION

There is much opportunity for choice in the design and implementation of technology. Although severe economic constraints and technical limits may be present in the early stages of a new technology or its application, more choice is generally possible in later stages. It is to the benefit of all concerned that users participate throughout the design and implementation of new technologies.

Although evidence is somewhat limited for the specific case of clerical workers and computer-based technologies, studies elsewhere of worker participation in the design and application of new equipment suggest that such participation can lead to substantial productivity improvements as well as to workers who are more satisfied because they have had a role in determining the direction of change. Women may be disadvantaged in their ability to participate because of their generally lesser technical background, lower-level positions, and lack of worker organizations, but they are well qualified to participate because of their first-hand knowledge of actual practice.

The effective implementation and use of new information technologies seem to require more active participation by workers at all levels. The flexibility of the new technologies makes possible many alternative organizations of work. For example, in some cases the role of supervisors and managers may be substantially reduced. In workplaces where participation in a variety of decisions affecting working conditions and job quality is common, participation in decisions about implementation approaches is also common. Such workplaces tend to exhibit high levels of acceptance of new technologies and high levels of worker satisfaction. Although evidence that such participation directly contributes to productivity increase is very scanty, the wise employer, eager to make the best use of new technologies, will involve all those likely to be affected. The importance of employment security to this cooperative process is clear; workers and managers will be unlikely to develop applications that eliminate their specific jobs (and presumably increase productivity) unless they believe their employment will continue in other capacities. Both workers and managers need to learn how to participate effectively in such collaborative processes.

The panel recommends that:

- manufacturers and designers of office equipment consult users about ergonomic and other features and about the development of technical standards; software developers and users be collaboratively involved in installation and applications; both managers and workers take part in evaluation and feedback to designers and manufacturers in order to promote ongoing improvement; and special attention be paid to women workers' knowledge of their jobs;

- employers involve everyone who will be affected by changes in equipment or work organization in the choice of system configuration and implementation;
- employers and workers educate themselves to participate in collective decision making.

MONITORING HEALTH CONCERNS

No permanent physiological damage has yet been shown to result from work with the new office technologies. Some concern about vision damage and possible radiation exposure has been expressed by those who work intensively with video display terminals (VDTs). A recent report by the National Research Council (1983) found no evidence of permanent damage to the eyes from video viewing, but it did find that eye strain and body fatigue are common. These negative effects are similar to those found among people who work with details on paper, but the rate of complaint is more frequent among VDT users. These effects can be largely eliminated by the proper application of current knowledge about sound ergonomic design, the use of appropriate furniture and lighting, and job design that includes variety and challenge. Despite good overall practices, some individual workers will experience clinical problems related to eye or body fatigue; these individual responses should be respected, and prudent employers should seek solutions to specific problems. If serious problems of a new sort develop (for example, if permanent health consequences emerge over a longer period), a more systematic response will be necessary.

Voluntary and cooperative actions cannot substitute for government protection in these circumstances. If new information technologies create demonstrable dangers to the health and safety of workers, present law will need to be examined to determine if it applies. New law or regulations may be required.

The panel recommends that:

- the federal government study the feasibility of conducting prospective epidemiological research on the health of workers exposed to VDTs (and related equipment) and fund such research if it is deemed appropriate.

DATA AND RESEARCH NEEDS

This report has frequently noted that currently available data are inadequate to fully understand the sources of change in the employment effects of technological developments, primarily because information about technology is not linked to information about workers.

Several steps can be taken to improve the situation. First, systematic comparative case studies and studies of firms in various industries, or of workers in several occupations, could be undertaken. Existing national data, and even data restricted to local labor markets, simply do not—and cannot—include the level of detail necessary to monitor the impact of technological change in jobs. Careful studies need to be designed that will follow both workers and jobs over time to determine how changes in technology affect the content of specific jobs, the division of labor between jobs, the organization of particular departments within enterprises, and the organization of the enterprise itself. Only through the accumulation of such comparative studies will understanding of these processes progress.

Second, the Bureau of Labor Statistics could use the results of such case studies to design special studies on the impact of technological change on the organization of work. The obvious limitation of case studies is their generalizability. Once it has been determined that a particular pattern of change is occurring, it is important to know how widespread it is, how many and what sorts of workers it affects, what kinds of firms are affected and in what way, and so forth. To answer such questions, national or labor market surveys, designed with explicit attention to the impact of technological change on work, could be helpful. The Bureau of Labor Statistics is the appropriate unit to establish a survey program that would be conducted by the Census Bureau. Such an arrangement would maximize the probability that data will be collected frequently from a representative sample of U.S. firms and workers. It might even be possible to develop a periodic survey instrument that tracks technological changes in firms and relates them to the employment characteristics of workers in the affected firms. Data on such worker characteristics as sex, race, ethnicity, and age should be collected because of likely differential effects. Because employers and workers may experience technological change differently, exploring survey instruments that query matched employer-employee "pairs" may be worthwhile.

A less ambitious but useful survey instrument is the Quality of Employment Survey. It is a survey of a national sample of workers that has been administered three times; the Department of Labor is currently considering the possibility of conducting it again in 1986. Adding questions about workers' experiences with specific sorts of technological changes, their access to training programs, their attitudes toward and use of new equipment, the extent of their involvement in decision making, and so on would probably generate some useful data. Whichever survey option is adopted, racial and ethnic minorities should be oversampled, and age cohorts should be large enough to permit valid comparisons across groups. Such studies and surveys will make possible improved understanding of the differential effects of technological change. (A fuller discussion of research and data needs can be found in the panel's interim report, Hunt and Hunt, 1985b.)

The panel recommends that government agencies support or conduct:

- case study research that systematically examines the employment impact of technological change in many occupations and industries;
- survey research, of a specific type of change, of a representative sample of firms (both employers and workers), to improve the ability to generalize about the effects of technological change;
- systematic research about the effects of technological change on the quality of work life and other human consequences of alternative applications of technology. Such studies should also examine the effects of alternative implementation approaches on productivity gains and the effects of worker participation in decision making; a comparative perspective would be especially useful for this subject.

EPILOGUE

During the past several years substantial structural changes have occurred in economic activity; the use of new information technologies has both contributed to these changes and been influenced by them. Employers and workers have had to cope with this period of change, and as the process continues to unfold, they will continue to do so. If all the panel's recommendations were adopted, workers and employers would benefit from effective implementation of new technologies and supportive government programs to assist with job transitions. The recommendations point to the relationships that exist between and among employers, workers, educational institutions, women's groups, equipment designers, manufacturers, and vendors. All these groups share significant interests in solving problems linked to technological change, although the distributive impact of costs, benefits, and possible solutions varies among and between them.

It must be noted, however, that the recommendations will not ensure sound economic performance and employment equity for women, no matter how broad the participation. The current period is one of economic uncertainty. In focusing on technological change, the panel has not addressed many policy issues that affect the economic future. Similarly, women's employment prospects are affected by much more than technological change and the policies discussed here that mitigate its worst or enhance its best effects. This period of transition can, however, prove a catalyst in addressing the broader issues, since, as we have seen, it is easier to adjust to technological change if economic growth is healthy and if affected workers have access to all parts of the economy. The healthier the economy, the better the status of women workers, the more likely it is that the current period of transition will be negotiated successfully by industry and by women workers. The panel believes that its recommendations provide a means of dealing effectively with the particular opportunities and problems that technological change poses for women workers. In its examination of the context in which technological change takes place and in which women work, this report may also suggest a broader consideration of economic and women's issues.

References

Abbott, Edith
    1910   *Women in Industry*. New York: Appleton.

Adler, Paul
    1984a   New Technologies, New Skills. Working Paper 9-784-076. Harvard Business School, Cambridge, Mass.
    1984b   Tools for resistance: workers can make automation their ally. *Dollars and Sense* 100 (October):7-8,17.

AFL-CIO Committee on the Evolution of Work
    1983   *The Future of Work*. Washington, D.C.: AFL-CIO.

Aguren, S., R. Hansson, and K. Karlsson
    1976   *The Impact of New Design on Work Organization*. Stockholm, Sweden: The Rationalization Council, SAF-LO.

Alexander, John J.
    1983   Case Study—Managing Automation. Paper presented at the National Executive Forum: Office Workstations in the Home. National Academy of Sciences, Washington, D.C., November 9-10.

American Nurses Association
    1985   *Facts About Nursing 84-85*. Kansas City, Mo.: American Nurses Association.

Anderson, Hobson Dewey, and Percy E. Davidson
    1940   *Occupational Trends in the United States*. Stanford, Calif.: Stanford University Press.

Appelbaum, Eileen
    1984   Technology and the Redesign of Work in the Insurance Industry. Project Report No. 84-A22. Institute for Research on Educational Finance and Governance, School of Education, Stanford University.
    1985   Alternative Work Schedules of Women. Paper prepared for the Panel on Technology and Women's Employment, Committee on Women's Employment and Related Social Issues, National Research Council, Washington, D.C. (July).

Armbruster, Albert
    1983   Ergonomic requirements. Pp. 169-189 in H. Otway and M. Peltu, eds., *New Office Technology: Human and Organizational Aspects*. Norwood, N.J.: Ablex.

Aronson, Sidney
  1977   Bell's electrical toy: what's the use? The sociology of early telephone usage. In Ithiel de Sola Pool, ed., *The Social Impact of the Telephone*. Cambridge, Mass.: MIT Press.

Attewell, Paul
  1985   The Automated Office: A Case Study. Unpublished manuscript. Department of Sociology, State University of New York, Stony Brook.
  In press   The deskilling controversy. *Work and Occupations*.

Attewell, Paul, and James Rule
  1984   Computing and organizations: what we know and what we don't know. *Communications of the ACM* 27:1184–1192.

Baker, Elizabeth Faulkner
  1964   *Technology and Women's Work*. New York: Columbia University Press.

Ball, Marion J., and Kathryn J. Hannah
  1984   *Using Computers in Nursing*. Reston, Va.: Reston Publishing Co.

Bancroft, N.
  1982   Productivity in the Office. Paper prepared for the Manufacturing Distribution and Control Business Group, Digital Equipment Corporation. Office Systems Consulting, Westminster, Mass.

Bancroft, N., E. Mumford, and B. Sontag
  No date   Participative Design: Successes and Problems. Unpublished paper. Digital Equipment Corporation, Maynard, Mass.

Baran, Barbara
  1985   The Technological Transformation of White-Collar Work: A Case Study of the Insurance Industry. Paper prepared for the Panel on Technology and Women's Employment, Committee on Women's Employment and Related Social Issues, National Research Council, Washington, D.C. (July).

Baran, Barbara, and Suzanne Teegarden
  1984   Women's Labor in the Office of the Future: Changes in the Occupational Structure of the Insurance Industry. Department of City and Regional Planning, University of California, Berkeley.

Barnowe, J.T., T.W. Mangione, and R.P. Quinn
  1973   Quality of employment indicators, occupational classifications, and demographic characteristics as predictors of job satisfaction. In R.P. Quinn and T.W. Mangione, eds., *The 1969–1970 Survey of Working Conditions: Chronicles of an Unfinished Enterprise*. Ann Arbor, Mich.: Institute for Social Research.

Baron, Ava
  1981   Woman's "Place" in Capitalist Production: A Study of Class Relations in the Nineteenth Century Newspaper Printing Industry. Ph.D. dissertation. New York University.

Bell, Carolyn Shaw
  1984   The Income Maintenance System Known as Employment. Working Paper No. 81. Department of Economics, Wellesley College.

Bem, Daryl, and H. McConnell
  1970   Testing the self-perception explanation of dissonance phenomena: on the salience of premanipulation attitudes. *Journal of Personality and Social Psychology* 14:23–31.

Benoit, Carmella, Alfred Cossette, and Prisco Cardillo
  1984   *L'incidence de la Machine á Traitement de Textes sur l'Emploi et le Travail*. Montreal, Quebec: Ministere du Travail.

Bermann, Tamar
  1985   Not only windmills: female service workers and new technologies. Pp. 231–248 in

REFERENCES

  A. Olerup, L. Schneider, and E. Morrod, eds., *Women, Work and Computerization: Opportunities and Disadvantages.* Amsterdam: Elsevier Science Publishers.

Bianchi, Suzanne M., and Daphne Spain
 1984 *American Women: Three Decades of Change.* Special Demographic Analyses. CDS-80-8. Washington, D.C.: U.S. Department of Commerce.

Bielby, William T., and James Baron
 1985 A woman's place is with other women: sex segregation within organizations. Pp. 27-55 in Barbara Reskin, ed., *Sex Segregation in the Workplace: Trends, Explanations, Remedies.* Committee on Women's Employment and Related Social Issues, National Research Council. Washington, D.C.: National Academy Press.

Bikson, Tora K.
 1986 Understanding the implementation of office technology. In Robert E. Kraut, ed., *Technology and the Transformation of White-Collar Work.* Hillsdale, N.J.: Lawrence Erlbaum.

Bikson, Tora K., and Barbara A. Gutek
 1983 *Advanced Office Systems: An Empirical Look at Utilization and Satisfaction.* Santa Monica, Calif.: Rand Corporation.

Bjorn-Anderson, Niels, and Dian Kjaergaard
 1986 Choices en route to the office of tomorrow. In Robert E. Kraut, ed., *Technology and the Transformation of White-Collar Work.* Hillsdale, N.J.: Lawrence Erlbaum.

Blau, Francine
 1977 *Equal Pay in the Office.* Lexington, Mass.: Lexington Books.

Blau, Francine D., and Marianne A. Ferber
 1986 *The Economics of Women, Men, and Work.* Englewood Cliffs, N.J.: Prentice-Hall.

Blomberg, Jeanette
 1986 Social interaction and office communication: effects on user evaluation of new technologies. In Robert E. Kraut, ed., *Technology and the Transformation of White-Collar Work.* Hillsdale, N.J.: Lawrence Erlbaum.

Bluestone, Barry, Patricia Hanna, Sarah Kuhn, and Laura Moore
 1981 *The Retail Revolution: Market Transformation, Investment, and Labor in the Modern Department Store.* Boston, Mass.: Auburn House Publishing Company.

Board on Telecommunications and Computer Applications
 1985 *Office Workstations in the Home.* Commission on Engineering and Technical Systems, National Research Council. Washington, D.C.: National Academy Press.

Bowen, Howard R., and Garth L. Mangum
 1966 *Automation and Economic Progress.* Summary report of the National Commission on Technology, Automation, and Economic Progress. Englewood Cliffs, N.J.: Prentice-Hall.

Braverman, Harry
 1974 *Labor and Monopoly Capital: The Degradation of Work in the Twentieth Century.* New York: Monthly Review Press.

Bucy, J. Fred
 1985 Computer sector profile. Pp. 46-78 in Anne G. Keatley, ed., *Technological Frontiers and Foreign Relations.* Report of the National Academy of Sciences, National Academy of Engineering, and Council on Foreign Relations. Washington, D.C.: National Academy Press.

Bureau of the Census
 No date *Projections of the Population of the United States by Age, Sex, and Race: 1983-2080.* Current Population Reports, Series P-25, No. 952. Washington, D.C.: U.S. Department of Commerce.

1964   *Census of Population 1950-1960. Detailed Characteristics.* Report PC (1C). Washington, D.C.: U.S. Department of Commerce.
1980   *Social Indicators III.* Washington, D.C.: U.S. Department of Commerce.
1983   *Statistical Abstract of the United States: 1984.* 104th edition. Washington, D.C.: U.S. Department of Commerce.
1984a   *Educational Attainment in the United States.* Current Population Reports, Series P-20, No. 390. Washington, D.C.: U.S. Department of Commerce.
1984b   *Census of Population 1970-1980. Detailed Occupation of the Experienced Civilian Labor Force by Sex for the United States and Regions.* Supplementary Report 80-S1-15. Washington, D.C.: U.S. Department of Commerce.
1985   *Statistical Abstract of the United States: 1986.* 106th edition. Washington, D.C.: U.S. Department of Commerce.

Bureau of Labor Statistics
1966   *The Impact of Office Automation in the Insurance Industry.* Bulletin 1468. Washington, D.C.: U.S. Department of Labor.
1980   *Perspectives on Working Women: A Data Book.* Bulletin 2080 (October). Washington, D.C.: U.S. Department of Labor.
1981   *Employment and Earnings,* 28 (January).
1985a   *Employment, Hours and Earnings, United States, 1904-84.* Vol. I. Bulletin 1312-12. Washington, D.C.: U.S. Department of Labor.
1985b   *Employment and Earnings,* 32 (January).
1986   *Employment and Earnings,* 33 (January).

Burns, William
1982   Changing corporate structure and technology in the retail food industry. Pp. 27-52 in Donald Kennedy, Charles Craypo, and Mary Lehman, eds., *Labor and Technology: Union Responses to Changing Environments.* University Park, Pa.: Department of Labor Studies, Pennsylvania State University.

Cain, Pamela
1985   Prospects for pay equity in a changing economy. Pp. 137-166 in Heidi I. Hartmann, ed., *Comparable Worth: New Directions for Research.* Committee on Women's Employment and Related Social Issues, National Research Council. Washington, D.C.: National Academy Press.

Carey, Max L., and Kim L. Hazelbaker
1986   Employment growth in the temporary help industry. *Monthly Labor Review* 109(4): 37-44.

Center for Career Research and Human Resources Management
1985   Results of the Study of the Introduction of Datapoint Terminals. Unpublished paper. Columbia University, New York.

Chamot, Dennis, and John L. Zalusky
1985   Use and misuse of workstations at home. Pp. 76-84 in *Office Workstations in the Home.* Report of the Board on Telecommunications and Computer Applications, Commission on Engineering and Technical Systems, National Research Council. Washington, D.C.: National Academy Press.

Child, John, Ray Loveridge, Janet Harvey, and Anne Spencer
1984   Microelectronics and the quality of employment in services. Pp. 163-190 in Pauline Marstrand, ed., *New Technology and the Future of Work and Skills.* London, England: Francis Pinter.

Christensen, Kathleen E.
1985   Women Who Work at Home: An Invisible Labor Force Made Visible. Center for Human

REFERENCES

Environments, Working Paper 90-PD-01. Graduate Center, City University of New York.

Cohn, Samuel
1985 *The Process of Occupational Sex-Typing: The Feminization of Clerical Labor in Great Britain, 1870–1936.* Philadelphia, Pa.: Temple University Press.

Cole, William
1986 Medical cognitive graphics. Pp. 91–95 in *Proceedings of CHI'86: Human Factors in Computing Systems*. New York: Association for Computing Machinery.

Commission of the European Communities
1984 *The Control of Frontiers: Workers and New Technology; Disclosure and Use of Company Information.* Final report presented to the Directorate General for Internal Market and Industrial Affairs, and Directorate General for Employment, Social Affairs, and Education, European Commission. Oxford, England: Ruskin College.

Committee on Science, Engineering, and Public Policy
1984 *High Schools and the Changing Workplace: The Employers' View.* Report of the Panel on Secondary School Education for the Changing Workplace, National Academy of Sciences, National Academy of Engineering, Institute of Medicine. Washington, D.C.: National Academy Press.

Communications Workers of America, AFL-CIO
1980 Contract negotiated between AT&T and the Communications Workers of America (CWA).

Corcoran, Mary, and Gregory J. Duncan
1979 Work history, labor force attachment, and earnings differences between the races and sexes. *Journal of Human Resources* 14(Winter):3–20.

Counte, Michael A., Kristen H. Kjerulff, Jeffrey C. Salloway, and Bruce C. Campbell
1983 Implementation of a medical information system: evaluation of adaptation. *Health Care Management Review* 8(2):25–33.

Cunningham, Nicholas, Carter Marshall, and Emily Glazer
1978 Telemedicine in pediatric primary care: favorable experience in nurse-staffed inner city clinic. *Journal of the American Medical Association* 240(25):2749–2751.

Daniels, Cynthia
1984 Between home and factory: homeworkers of New York, 1900–1914. Pp. 42–81 in Working Mothers and the State. Ph.D. dissertation. Department of Political Science, University of Massachusetts, Amherst.

Davies, Celia
1980 *Rewriting Nursing History.* London, England: Croom Helm, and Totowa, N.J.: Barnes and Noble.

Davies, Margery W.
1982 *Woman's Place Is at the Typewriter: Office Work and Office Workers, 1870–1930.* Philadelphia, Pa.: Temple University Press.

Denny, Michael, and Melvyn Fuss
1983 The effect of factor prices and technological change on the occupational demand for labor: evidence from Canadian telecommunications. *Journal of Human Resources* 18(2):161–176.

Department of Scientific and Industrial Research
1956 *Automation: A Report of Technical Trends and Their Impact on Management and Labor.* London, England. Unpublished.

de Sola Pool, Ithiel
1977a Introduction. Pp. 1–9 in Ithiel de Sola Pool, ed., *The Social Impact of the Telephone.* Cambridge, Mass.: MIT Press.

de Sola Pool, Ithiel, ed.
   1977b   *The Social Impact of the Telephone.* Cambridge, Mass.: MIT Press.

Dowling, Alan F.
   1980   Do hospital staff interfere with computer system implementation? *Health Care Management Review* 5(3):23-32.

Driscoll, James
   1980   Office Automation: The Dynamics of a Technological Boondoggle. Paper presented at the International Office Automation Symposium, Stanford University (March).

Dubnoff, Steven
   1978   Inter-Occupational Shifts and Changes in the Quality of Work in the American Economy, 1900-1970. Paper presented at the annual meeting of the Society for the Study of Social Problems, San Francisco, Calif.

Dumais, Susan, Susan Koch, and Robert Kraut
   1986   Implementing Service Order Computer Systems in a Telephone Company. Unpublished raw data. Bell Communications Research, Inc., Morristown, N.J.

Duncan, Gregory J., and Saul Hoffman
   1979   On-the-job training and earnings differences by race and sex. *Review of Economics and Statistics* 61(4):594-603.

Erickson, E.
   1934   *The Employment of Women in Offices.* Bulletin 120. Washington, D.C.: Women's Bureau, U.S. Department of Labor.

Ernst, Martin
   1982   The mechanization of commerce. *Scientific American* 247(3):132-147.

Evans, John
   1983   Negotiating technological change. Pp. 152-168 in H. Otway and M. Peltu, eds., *New Office Technology: Human and Organizational Aspects.* Hillsdale, N.J.: Ablex.

Fagerhaugh, Shizuko, Anselm Strauss, Barbara Suczek, and Carolyn Wiener
   1980   The impact of technology on patients, providers, and care patterns. *Nursing Outlook* November:666-672.

Feldberg, Roslyn
   1986   Technology and the transformation of clerical work. In Robert E. Kraut, ed., *Technology and the Transformation of White-Collar Work.* Hillsdale, N.J.: Lawrence Erlbaum.

Feldberg, Roslyn L., and Evelyn N. Glenn
   1979   Male and female: job vs. gender models in the sociology of work. *Social Problems* 26(5):524-538.
   1983   Technology and work degradation: effects of office automation on women clerical workers. Chapter 4 in Joan Rothschild, ed., *Machina ex Dea: Feminist Perspectives on Technology.* New York: Pergamon Press.

Flaim, Paul, and Ellen Sehgal
   1985   Displaced workers of 1973-1983: how well have they fared? *Monthly Labor Review* 108(6):3-16.

Fleming, John
   1985   Observations on Technology in the Retail Trade. Unpublished paper. Department of Psychology, Princeton University.

Fox, John
   1977   Medical computing and the user. *International Journal of Man-Machine Studies* 9:669-686.

Freeman, Richard B.
   1976   *The Overeducated American.* New York: Academic Press.

Fullerton, Howard N., Jr.
1985   The 1995 labor force: BLS' latest projections. *Monthly Labor Review* 108(11):17-25.
Ginzberg, Eli
1982   The mechanization of work. *Scientific American* 247(3):66-76.
Giuliano, Vincent E.
1982   The mechanization of office work. *Scientific American* 247(3):149-164.
Goldin, Claudia
1985   Women's Employment and Technological Change: An Historical Perspective. Paper prepared for the Panel on Technology and Women's Employment, Committee on Women's Employment and Related Social Issues, National Research Council, Washington, D.C. (October).
Gorlin, Harriet, and Lawrence Schein
1984   Innovations in Managing Human Resources. Research Report No. 849. The Conference Board, New York.
Gottmann, Jean
1977   Megalopolis and antipolis: the telephone and the structure of the city. Pp. 303-317 in Ithiel de Sola Pool, ed., *The Social Impact of the Telephone*. Cambridge, Mass.: MIT Press.
Grobe, Susan J.
1984   Nursing education module authoring system—NEMAS. *Journal of Educational Technology Systems* 13(2):83-89.
Gurstein, Michael, and Ferdinand Faulkner
1985   Office Automation and Organizational Culture: A Case Study of Two Implementations of an Automatic Call Distributor. Paper presented at the Computer and Society Conference (June 21), Rochester, N.Y.
Gurwitz, Aaron S., and Julie N. Rappaport
1984– Structural change and slower employment growth in the financial services sector. *Federal Reserve Bank of New York Review* 9(4):39-45.
1985
Gutek, Barbara, and Tora K. Bikson
1985   Differential experiences of men and women in computerized offices. *Sex Roles* 13(3-4):123-136.
Gutek, Barbara, Tora K. Bikson, and Don Mankin
1984   Individual and organizational consequences of computer-based office information technology. Pp. 231-254 in S. Oskamp, ed., *Applied Social Psychology Annual*. 5th edition. Beverly Hills, Calif.: Sage Publications.
Guzda, Henry
1984   Industrial democracy: made in the U.S.A. *Monthly Labor Review* 107(5):26-33.
Hackman, J. Richard, and Greg Oldham
1980   *Work Redesign*. Santa Monica, Calif.: Goodyear Publishing Company.
Hackman, J. Richard, Jane L. Pearce, and Jane C. Wolfe
1978   Effects of change in job characteristics on work attitudes and behaviors: a naturally occurring quasi-experiment. *Organizational Behavior and Human Performance* 21: 289-304.
Hartmann, Heidi I.
1981   The family as the locus of gender, class and political struggle: the example of housework. *Signs: Journal of Women in Culture and Society* 6(3):366-394.
Helander, M., and B. Rupp
1984   An overview of standards and guidelines for visual display terminals. *Applied Ergonomics* 3:185-195.

Helfgott, Roy B.
    1966   EDP and the office work force. *Industrial and Labor Relations Review* 19(July):503–516.

Henskes, D.T., and H.E. Kronick
    1974   Operator acceptance of data entry devices in patient care areas of a hospital. Pp. 639–643 in *Medinfo*. Amsterdam, Holland: North-Holland Publishing Company.

Herman, G., K. Owens, and H. Schottland
    1979   Service Representative Questionnaire Study. Technical Memorandum No. 79-3431-2. Bell Laboratories, Murray Hill, N.J.

Hirschhorn, Larry
    No date   Office Automation and the Entry Level Job. A Concept Paper. Management and Behavioral Science Center, Wharton School, University of Pennsylvania.

Honeywell Corporation
    1983   *National Survey on Office Automation and the Workplace*. Minneapolis, Minn.: Honeywell Corp.

Horowitz, Morris, and Irwin Herrenstadt
    1966   Changes in skill requirements of occupations in selected industries. Pp. 223–287 in *The Employment Impact of Technological Change: Technology and the American Economy*. Report of the U.S. National Commission on Technology, Automation, and Economic Progress. Volume 2. Washington, D.C.: U.S. Government Printing Office.

Howard, Robert
    1985   UTOPIA: where workers craft new technology. *Technology Review* 88:43–49.

Howard, Robert, and Leslie Schneider
    1985   Worker Participation in Technological Change: Interests, Influence, and Scope. Paper prepared for the Panel on Technology and Women's Employment, Committee on Women's Employment and Related Social Issues, National Research Council, Washington, D.C. (February).

Hunt, H. Allan, and Timothy L. Hunt
    1985a   Clerical Employment and Technological Change: A Review of Recent Trends and Projections. Paper prepared for the Panel on Technology and Women's Employment, Committee on Women's Employment and Related Social Issues, National Research Council, Washington, D.C. (July).
    1985b   An assessment of data sources to study the employment effects of technological change. Pp. 1–116 in *Technology and Employment Effects*. Interim report of the Panel on Technology and Women's Employment, Committee on Women's Employment and Related Social Issues, National Research Council. Washington, D.C.: National Academy Press.

Iacono, Suzanne, and Rob Kling
    1986   Changing office technologies and transformations of clerical work. In Robert E. Kraut, ed., *Technology and the Transformation of White-Collar Work*. Hillsdale, N.J.: Lawrence Erlbaum.

Jackson, Robert Max
    1984   *The Formation of Craft Labor Markets*. Orlando, Fla.: Academic Press.

Jacobson, Sharol F.
    1983   The contexts of nurses' stress. Pp. 49–60 in S.F. Jacobson and H.M. McGrath, eds., *Nurses Under Stress*. New York: John Wiley & Sons.

Jacobson, Sharol F., and H. Marie McGrath, eds.
    1983   *Nurses Under Stress*. New York: John Wiley & Sons.

Johnson, Bonnie McD., and Ronald E. Rice
    1983   Policy Implications in Implementing Office Systems Technology. Paper presented at the 11th Annual Telecommunications Research and Policy Conference, Annapolis, Md. (April).

REFERENCES

Johnson, Bonnie, James C. Taylor, Dennis R. Smith, and Timothy R. Cline
No   Innovation in Word Processing: Report of a Maturing Technology. NSF project ISI
date   8110791. National Science Foundation, Washington, D.C.

Katz, Daniel, and Robert Kahn
1978   *The Social Psychology of Organizations.* 2nd edition. New York: John Wiley & Sons.

Katz, Elihu, and Paul F. Lazarsfeld
1955   *Personal Influence.* Glencoe, Ill.: Free Press.

Katzell, R., and R. Guzzo
1983   Psychological approaches to productivity improvement. *American Psychologist* 38:468-472.

Kelly Services
1984   *The Kelly Report on People in the Electronic Office. Volume 3: The Secretary's Role.* Prepared by Research and Forecasts, Inc., New York.

Kling, Robert
1978   The Impacts of Computing on the Work of Managers, Data Analysts and Clerks. Working paper. Public Policy Research Organization, University of California, Irvine.
1980   Social analyses of computing: theoretical perspectives in recent empirical research. *Computing Surveys* 12:61-110.

Kling, Robert, and Suzanne Iacono
1984   Computing as an occasion for social control. *Journal of Social Issues* 40:77-97.

Kohl, George
1986   Technological change in telecommunications: its impact on work and workers. Chapter 15 in Richard Gordon, ed., *Microelectronics in Transition: Industrial Transformation and Social Change.* Totowa, N.J.: Ablex.

Kraemer, Kenneth, and James Danziger
1982   Computers and Control in the Work Environment. Public Policy Research Organization, University of California, Irvine.

Kraft, Philip
1977   *Programmers and Managers.* New York: Springer-Verlag.
1985   A review of empirical studies of the consequences of technological change on work and workers in the United States. Pp. 117-150 in *Technology and Employment Effects.* Interim report of the Panel on Technology and Women's Employment, Committee on Women's Employment and Related Social Issues, National Research Council. Washington, D.C.: National Academy Press.

Kraft, Philip, and Steven Dubnoff
1986   Job content, fragmentation and control in computer software work. *Industrial Relations* 25(Spring):184-196.

Kraut, Robert E.
In   Telework as a work-style innovation. *Information and Behavior.* Volume 2.
press

Kraut, Robert E., and Patricia Grambsch
1985   Prophecy by Analogy: Potential Causes for and Consequences of Electronic Homework. Paper presented at the Computer and Society Conference, June 21, Rochester, N.Y.

Kusterer, Ken
1979   *Know-How on the Job: The Important Working Knowledge of "Unskilled" Workers.* Boulder, Colo.: Westview Press.

Langlotz, Curtis, and Edward Shortliffe
1983   Adapting a consultation system to critique user plans. *International Journal of Man-Machine Studies* 19:479-496.

Laws, Judith Long
  1976 The Bell telephone system: a case study. Pp. 157–178 in Phyllis A. Wallace, ed., *Equal Employment Opportunity and the AT&T Case*. Cambridge, Mass.: MIT Press.
Leffingwell, W.H.
  1925 *Office Management: Principles and Practice*. Chicago and New York: A.W. Shaw.
Leontief, Wassily
  1983 National perspective: the definition of problems and opportunities. Pp. 3–7 in *The Long-Term Impact of Technology on Employment and Unemployment*. Report of a National Academy of Engineering Symposium. Washington, D.C.: National Academy Press.
Leontief, Wassily, and Faye Duchin
  1984 The Impacts of Automation on Employment, 1963–2000. New York: Institute for Economic Analysis, New York University.
Lievrouw, Leah
  1984 Hospital Information Technology and Work Behavior. Paper prepared for the annual meeting of the International Communication Association, May. Annenberg School of Communications, University of Southern California, Los Angeles.
Light, Nancy
  1983 Computers in nursing practice: on-line care programs. *Computers in Nursing* 1(6):4.
Lipset, Seymour Martin, Martin Trow, and James Coleman
  1956 *Union Democracy: The Internal Politics of the International Typographical Union*. Glencoe, Ill.: Free Press.
Locke, Edwin
  1976 The nature and causes of job satisfaction. Pp. 1297–1349 in M. Dunnette, ed., *Handbook of Industrial and Organizational Psychology*. Chicago, Ill.: Rand McNally.
Lockwood, David
  1958 *The Blackcoated Worker: A Study in Class Consciousness*. London, England: George Allen and Unwin Ltd.
Machung, Anne
  1983 Turning secretaries into word processors: some fiction and a fact or two. Pp. 119–123 in D. Marschall and J. Gregory, eds., *Office Automation: Jekyll or Hyde?* Cleveland, Ohio: Working Women Education Fund.
  1984 Word processing: forward for business, backward for women. Pp. 124–139 in Karen Sacks and Dorothy Remy, eds., *My Troubles Are Going To Have Trouble with Me: Everyday Trials and Triumphs of Women Workers*. New Brunswick, N.J.: Rutgers University Press.
Maddox, Brenda
  1977 Women and the switchboard. Pp. 262–280 in Ithiel de Sola Pool, ed., *The Social Impact of the Telephone*. Cambridge, Mass.: MIT Press.
Maidique, Modesto A., and Robert M. Hayes
  1984 The art of high-technology management. *Sloan Management Review* 25(Winter):17–31.
Maidique, Modesto A., and Billie Joe Zirger
  1984 A study of successes and failures in product innovations: the case of the U.S. electronics industry. *IEEE Transactions* Nov.(EM)-31:192–203.
Malveaux, Julianne
  1982 Recent Trends in Occupational Segregation by Race and Sex. Paper presented at the Workshop on Job Segregation by Sex, Committee on Women's Employment and Related Social Issues, National Research Council, Washington, D.C.
Mansfield, Edwin
  1966 Technological change: measurement, determinants, and diffusion. Pp. 97–131 in *The*

Employment Impact of Technological Change: Technology and the American Economy. Report of the U.S. National Commission on Technology, Automation, and Economic Progress. Volume 2. Washington, D.C.: U.S. Government Printing Office.

Mark, Jerome
 1979 Measuring the effects of technological change. Pp. 18–21 in Dennis Chamot and Joan M. Baggett, eds., *Silicon, Satellites and Robots: The Impacts of Technological Change on the Workplace.* Washington, D.C.: AFL-CIO.

Marschall, Daniel, and Judith Gregory, eds.
 1983 *Office Automation: Jekyll or Hyde?* Cleveland, Ohio: Working Women Education Fund.

Matteis, R.J.
 1979 The new back office focuses on customer service. *Harvard Business Review* 57:146–159.

McDavid, Mary
 1985 U.S. Army: prototype program for professionals. Pp. 24–32 in *Office Workstations in the Home.* Report of the Board on Telecommunications and Computer Applications, Commission on Engineering and Technical Systems, National Research Council. Washington, D.C.: National Academy Press.

McNeill, Donna G.
 1979 Developing the complete computer-based information system. *Journal of Nursing Administration* 6(11):34–46.

Medical Informatics Europe
 1982 *Proceedings of the Fourth Congress of the European Federation of Medical Informatics* (Dublin, Ireland, March 21–25). R.R. O'Moore, B. Barber, P.L. Reichertz, and F. Roger, eds. New York: Springer-Verlag.
 1984 *Proceedings of the Sixth Congress of the European Federation of Medical Informatics* (Brussels, Belgium, September 10–13). F. Roger, J.L. Willems, R. O'Moore, and B. Barber, eds. New York: Springer-Verlag.

Mills, C. Wright
 1956 *White Collar: The American Middle Classes.* New York: Oxford [Originally published 1951].

Minolta Corporation
 1983 *The Evolving Role of the Secretary in the Information Age.* Ramsey, N.J.: Minolta Corporation.

Mirvis, Philip, and Edward Lawler
 1977 Measuring the financial impact of employee attitudes. *Journal of Applied Psychology* 62:1–8.

Morse, N., and E. Reimer
 1956 The experimental change of a major organizational variable. *Journal of Abnormal and Social Psychology* 20:191–198.

Mumford, Enid
 1981 *Values, Technology and Work.* The Hague, Netherlands: Martinus Nijhoff Publishers.
 1983 Designing Participatively. Manchester Business School, Manchester, England. Unpublished.

Mumford, Enid, and M. Weir
 1979 *Computer Systems in Work Design: The ETHICS Method: Effective Technical and Human Implementation of Computer-Based Systems.* New York: Halsted Press.

Murolo, Priscilla
 1986 White-collar women and the rationalization of clerical work: the Aetna Life Insurance Company, 1910–1930. In Robert E. Kraut, ed., *Technology and the Transformation of White-Collar Work.* Hillsdale, N.J.: Lawrence Erlbaum.

Murphree, Mary C.
1983 Dealing with technological discrimination. *Secretary Speakout '83: The Professional Secretary's New Identity in the Information Age.* Report of a conference March 9-11, Boston, Mass. Professional Secretaries International.
1984 Brave new office: the changing world of the legal secretary. Pp. 140-159 in Karen Sacks and Dorothy Remy, eds., *My Troubles Are Going to Have Trouble with Me: Everyday Trials and Triumphs of Women Workers.* New Brunswick, N.J.: Rutgers University Press.
1985 New Technology and Office Tradition: The Not-So-Changing World of the Secretary. Paper prepared for the Panel on Technology and Women's Employment, Committee on Women's Employment and Related Social Issues, National Research Council, Washington, D.C. (July).

National Center for Education Statistics
1984 The Condition of Education: A Statistical Report. Washington, D.C.: U.S. Department of Education.

National League for Nursing
1985 Table on Licensed Practical Nurse Population, by Sex, Racial/Ethnic Background, and Age Group. National Sample Survey. National Technical Information Service, U.S. Department of Commerce, Springfield, Va.

National Research Council
1979 *Measurement and Interpretation of Productivity.* Report of the Panel to Review Productivity Statistics, Committee on National Statistics. Washington, D.C.: National Academy of Sciences.
1983 *Video Displays, Work, and Vision.* Report of the Panel on Impact of Video Viewing on Vision of Workers, Committee on Vision. Washington, D.C.: National Academy Press.
1986 *Human Resource Practices for Implementing Advanced Manufacturing Technology.* Report of the Committee on the Effective Implementation of Advanced Manufacturing Technologies, Manufacturing Studies Board. Washington, D.C.: National Academy Press.

National Science Foundation, Office of Scientific and Engineering Personnel and Education
No *Science and Engineering Education: Data and Information.* Report prepared for the
date National Science Board on Precollege Education in Mathematics, Science, and Technology. Washington, D.C.: National Science Foundation.

Nelson, Kristen
1983 Back Offices and Female Labor Markets: Office Suburbanization in the San Francisco Bay Area. Ph.D. dissertation. University of California, Berkeley.

9-to-5
1984a *The 9-to-5 National Survey on Women and Stress* (with Office Automation Addendum). Cleveland, Ohio: 9-to-5, National Association of Working Women.
1984b *9-to-5 Campaign on VDT Risks: Analysis of VDT Operator Questionnaires.* Cleveland, Ohio: 9-to-5, National Association of Working Women.
1985 *Hidden Victims: Clerical Workers, Automation, and the Changing Economy.* Cleveland, Ohio: 9-to-5, National Association of Working Women.

Nisbett, Richard, and Timothy Wilson
1977 Telling more than we can know: verbal reports on mental processes. *Psychological Review* 84:231-259.

Noble, David
1984 Tools of oppression. *Dollars and Sense* 100 (October):6, 17.

Noble, Kenneth B.
1986 Study Finds 60% of 11 Million Who Lost Jobs Got New Ones. *New York Times* (February 7):A1 and A15.

REFERENCES

Norstedt, J.P., and S. Aguren
  1973 *The Saab-Scania Report.* Stockholm, Sweden: Swedish Employers' Confederation.

Northrup, Herbert R., and John A. Larson
  1979 *The Impact of the AT&T-EEO Consent Decrees.* Labor Relations and Public Policy Series No. 20 (November). Wharton School Industrial Research Unit, University of Pennsylvania.

Noyelle, Thierry J.
  1985 The New Technology and the New Economy: Implications for Equal Employment Opportunity. Paper prepared for the Panel on Technology and Women's Employment, Committee on Women's Employment and Related Social Issues, National Research Council, Washington, D.C. (July).

Office of Economic Growth and Employment Projections
  1981 Projected Occupational Staffing Patterns of Industries. OES Technical Paper No. 2. Washington, D.C.: U.S. Department of Labor.

Office of Technology Assessment, U.S. Congress
  1985 *Automation of America's Offices, 1985-2000.* OTA-CIT-287. Washington, D.C.: U.S. Government Printing Office.
  1986 *Technology and Structural Unemployment: Reemploying Displaced Adults.* OTA-ITE-250. Washington, D.C.: U.S. Government Printing Office.

Olson, Margrethe
  1983 Remote office work: changing work patterns in space and time. *Communications of the ACM* 26:182-187.
  1986 Telework: practical experience and future prospects. In Robert E. Kraut, ed., *Technology and the Transformation of White-Collar Work.* Hillsdale, N.J.: Lawrence Erlbaum.

Olson, Margrethe, and S. Primps
  1984 Working at home with computers: work and nonwork issues. *Journal of Social Issues* 40:97-112.

Ontario Retail Council, United Food and Commercial Workers Union
  1981 *Preliminary Investigation of Ring-and-Bag Check-Out Systems.* Ontario, Canada: Ontario Retail Council, United Food and Commercial Workers Union.

Oppenheimer, Valerie Kincade
  1970 *The Female Labor Force in the United States.* Population Monograph Series, No. 5. Berkeley: University of California, Institute of International Studies.

Osterman, Paul
  1984 White-collar internal labor markets. Chapter 6 in Paul Osterman, ed., *Internal Labor Markets.* Cambridge, Mass.: MIT Press.

Panko, Raymond
  1984 Office work. *Office: Technology and People* 2:205-238.

Podgursky, Michael
  1984 Sources of secular increases in the unemployment rate, 1969-82. *Monthly Labor Review* 107(7):19-25.
  1986 Job Displacement and Labor Market Adjustment: Evidence from the Displaced Worker Survey. Paper prepared for the Panel on Technology and Employment, Committee on Science, Engineering, and Public Policy, National Academy of Sciences, Washington, D.C.

Podgursky, Michael, and Paul Swaim
  1986 Labor Market Adjustment and Job Displacement: Evidence from the January 1984 Displaced Worker Survey. Final Report to the Bureau of International Affairs, U.S. Department of Labor (January).

Poppel, Harvey
  1982 Who needs the office of the future? *Harvard Business Review* 60(6):146-155.

Pratt, Jane H.
   1984   Home teleworking: a study of its pioneers. *Technological Forecasting and Social Change* 25:1–14.

Presser, Harriet B., and Wendy Baldwin
   1980   Child care as a constraint on employment: prevalence, correlates, and bearing on the work and fertility nexus. *American Journal of Sociology* 85(March):1202–1213.

Quinn, Robert, and Graham Staines
   1977   *The 1977 Quality of Employment Survey.* Ann Arbor, Mich.: Institute for Social Research.

Reskin, Barbara F., and Heidi I. Hartmann, eds.
   1986   *Women's Work, Men's Work: Sex Segregation on the Job.* Report of the Committee on Women's Employment and Related Social Issues, National Research Council. Washington, D.C.: National Academy Press.

Rhee, H.A.
   1968   *Office Automation in Social Perspective: The Progress and Social Implications of Electronic Data Processing.* Oxford, England: Basil Blackwell.

Rice, Ronald, Bonnie Johnson, Deborah Kowal, and Charles Feltman
   1983   The Survival of the Fittest: Organizational Design and the Structuring of Word Processing. Paper presented at the meeting of the Academy of Management, Dallas, Tex. (August).

Richer, Mark
   1986   GUIDON-2: A Knowledge-Based Learning Environment for Diagnosis. Paper and demonstration presented at CHI'86: Human Factors in Computing Systems. April 1986, Boston, Mass.

Rogers, T., and N. Friedman
   1980   *Printers Face Automation.* Lexington, Mass.: Lexington Books.

Rosenberg, Nathan
   1976   *Perspectives on Technology.* Cambridge, England: Cambridge University Press.
   1983   Testimony before Joint Subcommittee of the House Science and Technology Committee and the House Budget Committee. Pp. 29–43 in *Technology and Employment,* Joint Hearings Before the Subcommittee on Science, Research, and Technology of the Committee on Science and the Task Force on Education and Employment of the Committee on the Budget, U.S. House of Representatives, Serial No. TF4-4. Washington, D.C.: U.S. Government Printing Office.

Ross, Ian M.
   1985   Telecommunications. Pp. 22–45 in Anne G. Keatley, ed., *Technological Frontiers and Foreign Relations.* Report of the National Academy of Sciences, National Academy of Engineering, and Council on Foreign Relations. Washington, D.C.: National Academy Press.

Rumberger, Russell W.
   1981   The changing skill requirements of jobs in the U.S. economy. *Industrial and Labor Relations Review* 34:578–590.
   1984   The Potential Impact of Technology on the Skill Requirements of Future Jobs. Project report No. 84-A24. Institute for Research on Educational Finance and Governance, School of Education, Stanford University.

Ryan, Sheila A.
   1985   An expert system for nursing practice. *Computers in Nursing* 3(2):77–84.

Rytina, Nancy F., and Suzanne M. Bianchi
   1984   Occupational reclassification and changes in distribution by gender. *Monthly Labor Review* 107(3):11–17.

REFERENCES

Sadler, Philip
1981 Welcome back to the "automation" debate. Pp. 290–296 in Tom Forester, ed., *The Microelectronics Revolution: The Complete Guide to the New Technology and Its Impact on Society.* Cambridge, Mass.: MIT Press.

Salmans, Sandra
1982 The debate over the electronic office. *New York Times* (November 14)Section 6:B32-B37.

Schmitt, Roland S.
1983 Technological trends. Pp. 8–12 in *The Long-Term Impact of Technology on Employment and Unemployment.* Report of a National Academy of Engineering Symposium. Washington, D.C.: National Academy Press.

Scott, Joan W.
1982 The mechanization of women's work. *Scientific American* 247(3):186–187.

Sekscenski, Edward
1984 *The Health Services Industry in the United States: Trends in Employment from 1970 to 1983, with Projections to 1995.* Publication Nos. 84-2 and 84-3 (August). Department for Professional Employees. Washington, D.C.: AFL-CIO.

Service Employees International Union, AFL-CIO
1984 Contract Between the Service Employees International Union (SEIU) and the Equitable Life Assurance Society of the United States, Covering Employees in the Syracuse, New York, Claims Benefit Office. Syracuse, N.Y.: Service Employees International Union.

Silvestri, George T., and John M. Lukasiewicz
1985 Occupational employment projections: the 1984–95 outlook. *Monthly Labor Review* 108(11):42–59.

Silvestri, George T., John M. Lukasiewicz, and Marcus E. Einstein
1983 Occupational employment projections through 1995. *Monthly Labor Review* 106(11):37–49.

Simon, Herbert A.
1977 What computers mean for man and society. *Science* 195:1186–1191.

Smith, M.J.
1984 Human factors issues in VDT use: environmental and workstation design considerations. *IEEE Computer Graphics and Applications* 4:56–63.

Smith, Shirley J.
1985 Revised worklife tables reflect 1979–80 experience. *Monthly Labor Review* 108(8):23–30.

Smith, James P., and Michael P. Ward
1984 *Women's Wages and Work in the Twentieth Century.* Santa Monica, Calif.: The Rand Corporation.

Spenner, Kenneth
1979 Temporal changes in work content. *American Sociological Review* 69:965–975.

Spinrad, R.J.
1982 Office automation. *Science* 215:808–813.

Startsman, Terry S., and Robert E. Robinson
1972 The attitudes of medical and paramedical personnel towards computers. *Computers and Biomedical Research* 5:218–227.

Strassman, Paul
1980 The office of the future: information management for the new age. *Technology Review* 82:54–65.

Strober, Myra, and Carolyn Arnold
1985 Integrated Circuits/Segregated Labor: Women in Three Computer-Related Occupa-

tions. Paper prepared for the Panel on Technology and Women's Employment, Committee on Women's Employment and Related Social Issues, National Research Council, Washington, D.C. (July).

Strom, Sharon Hartman
 1985 Technology, the Office, and the Changing Sexual Division of Labor, 1910-1940. Paper prepared for the Panel on Technology and Women's Employment, Committee on Women's Employment and Related Social Issues, National Research Council, Washington, D.C. (July).

Suchman, Lucy, and Eleanor Wynn
 1984 Procedures and problems in the office. *Office: Technology and People* 2(2):133-154.

Sullivan, Teresa A.
 1978 Racial/ethnic differences in labor force participation. Pp. 165-187 in F.D. Bean and W.P. Frisbie, eds., *The Demography of Racial and Ethnic Groups*. New York: Academic Press.

Taylor, James C.
 1986 Job design and the quality of working life. In Robert E. Kraut, ed., *Technology and the Transformation of White-Collar Work*. Hillsdale, N.J.: Lawrence Erlbaum.

Thoite, Peggy
 1983 Multiple identities and psychological well-being: a reformulation and test of the social isolation hypothesis. *American Sociological Review* 48:174-187.

Thorsrud, E., B.S. Sorensen, and B. Gustavsen
 1976 Sociotechnical approach to industrial democracy in Norway. In R. Dubin, ed., *Handbook of Work, Organization, and Society*. Chicago, Ill.: Rand McNally.

Tilly, Charles, and Louise Tilly
 1985 Historical Studies at the New School. Working Paper No. 1. Center for Studies of Social Change, New School for Social Research.

Treiman, Donald J., and Kermit Terrell
 1975 Sex and the process of status attainment: a comparison of working women and men. *American Sociological Review* 40(April):174-200.

Turner, Jon
 1984 Computer mediated work: the interplay between technology and structured jobs. *Communications of the ACM* 27:1200-1210.

U.S. Department of Labor
 1974 *Job Satisfaction: Is There a Trend?* Manpower Research Monograph No. 30. GPO Stock Number 2900-00195. Washington, D.C.: U.S. Department of Labor.

U.S. National Commission on Technology, Automation, and Economic Progress
 1966 *The Employment Impact of Technological Change: Technology and the American Economy*. Report of the U.S. National Commission on Technology, Automation, and Economic Progress, Volume 1, Serial No. 0-788-561. Washington, D.C.: U.S. Government Printing Office.

Von Hippel, Eric
 1978 Users as innovators. *Technology Review* 80:31-39.

Wallersteiner, Ulrika
 1981 An outline of ergonomic research as applied to the design of supermarket cashier work stations. *Labour Research Bulletin* December:19-26.

Walton, Richard E.
 1975 The diffusion of new work structures: explaining why success didn't take. *Organizational Dynamics* 3(3):2-22.

Watson, R.J.
 1974 Medical staff response to a medical information system with direct physician-computer interface. Pp. 299-302 in *Medinfo*. Amsterdam, Holland: North-Holland Publishing Co.

Westin, Alan
   1985   Good User Practices in the Application of Office Systems Technology to Clerical Work, with Special Attention to Women's Equality and Workplace Quality Issues. Paper prepared for the Panel on Technology and Women's Employment, Committee on Women's Employment and Related Social Issues, National Research Council, Washington, D.C. (July).

Westin, Alan, H. Schweder, M. Baker, and S. Lehman
   1985   *The Changing Workplace: A Guide to Managing the People, Organizational and Regulatory Aspects of Office Technology.* Westchester, N.Y.: Knowledge Industries.

White, Julie
   1985   *Trouble in Store? The Impact of Microelectronics in the Retail Trade.* Ottawa, Ont.: Women's Bureau, Labor Canada.

Women's Bureau, U.S. Department of Labor
   1983   *Handbook on Women Workers.* Bulletin 298. Washington, D.C.: U.S. Department of Labor.
   1985   Earnings difference between women and men workers. *Facts on U.S. Working Women.* 85-7 (July). Washington, D.C.: U.S. Department of Labor.

Yin, Robert K., and Gwendolyn B. Moore
   1984   *Planning and Implementing an Automated Office.* Washington, D.C.: COSMOS Corporation.

Yin, Robert K., and J. Lynne White
   1984   Microcomputer Implementation in Schools. Washington, D.C.: COSMOS Corporation.

Zuboff, Shoshanah
   1982   Problems of symbolic toil. *Dissent* 29 (Winter): 51-61.

Biographical Sketches of Panel Members and Staff

TAMAR D. BERMANN is a senior researcher at the Work Research Institutes in Oslo, Norway, and teaches social policy at the University of Oslo. Previously she was senior lecturer at the Roskilde University Center in Denmark. She has also served as a research fellow at the Max Planck Research Centre for Psychopathology and Psychotherapy in Munich, West Germany, and as a teacher at the University of Munich. She has served on technology research committees in Denmark, Sweden, and Norway; worked with the women's studies center of the Norwegian Research Council; and participated in the research program of the Council of Europe. Her action-oriented research focuses on learning and knowledge acquisition in connection with organizational and technological change, particularly in journalism, nursing, librarianship, and office work. Her scholarly interests include epistemology and the history of science. She has a Ph.D. in sociology from the University of Munich and a Ph.D. in philosophy from the University of Vienna. She is a native of Vienna, Austria.

FRANCINE D. BLAU is professor of economics and labor and industrial relations at the University of Illinois at Urbana-Champaign. She is a former vice-president of the Midwest Economics Association and was a member of the American Economic Association Committee on the Status of Women in the Economics Profession. She currently serves on the National Research Council's Panel on Pay Equity Research. Her research has centered on women's economic status and discrimination against women and minorities. She has also studied a variety of issues related to immigration, job search and turnover, union impact, and racial differences in wealth. She has served as an expert witness in employment discrimination cases and testified in Congress on the

Equal Rights Amendment and the economic status of women. She has a B.S. degree from Cornell University in industrial and labor relations and A.M. and Ph.D. degrees in economics from Harvard University.

DENNIS CHAMOT is associate director of the AFL-CIO's Department for Professional Employees. Prior to his work in organized labor, he was employed as a research chemist by E. I. du Pont de Nemours. He has for several years had a particular interest in the effects of new technologies on employment. He has served on numerous panels related to this issue, including the National Science Foundation Advisory Council; the Commission on Engineering and Technical Systems of the National Research Council (NRC); the Committee on the Education and Utilization of Engineers of the NRC; the Labor Research Advisory Council of the Bureau of Labor Statistics; and the Council of the American Chemical Society. He has B.S. and M.S. degrees from the Polytechnic Institute of Brooklyn (now the Polytechnic Institute of New York), a Ph.D. in chemistry from the University of Illinois, and an M.B.A. from the Wharton School of the University of Pennsylvania.

MARTIN L. ERNST is vice-president, advanced information technologies, at Arthur D. Little, Inc. His early career was devoted to operations research for the U.S. Navy and U.S. Air Force and later for Arthur D. Little. More recently he has concentrated on the impacts of technology on society, businesses, and other types of institutions, with particular emphasis on the influences of the new information technologies. He received a B.S. degree in physics from the Massachusetts Institute of Technology in 1941.

ROSLYN L. FELDBERG is an associate director of labor relations at the Massachusetts Nurses Association. Her background includes both scholarly and advocacy work in a wide variety of settings, from academia to public interest and labor groups. For several years her research interests have focused on clerical work. Recently she was a Radcliffe research scholar at the Murray Research Center at Radcliffe College, where she continued research on women clerical workers. Previously she was a coprincipal investigator for a major project on the impact of job conditions on women clerical workers, funded by the National Institute of Mental Health. She has also taught in Boston University's Department of Sociology and at the University of Aberdeen in Scotland and has been active in public policy debates on comparable worth, the interaction of work and family life, and the implications of technological change on women workers. She has also served as an adviser to the Office of Technology Assessment of the U.S Congress, 9-to-5, and the National Endowment for the Humanities. She has a B.A. from the University of Illinois and M.A. and Ph.D. degrees from the University of Michigan, all in sociology.

HEIDI I. HARTMANN is study director of the Committee on Women's Employment and Related Social Issues, the Panel on Technology and Women's Employment, and the Panel on Pay Equity Research at the National Research Council (NRC). She has edited or coedited a number of NRC reports on comparable worth and other women's employment issues. Previously she taught economics on the Graduate Faculty at the New School for Social Research. Her research has concentrated on employment issues related to women and minorities, particularly discrimination and internal labor markets; women's economic independence; and political economy and feminist theory. She has a B.A from Swarthmore College and M.Ph. and Ph.D. degrees from Yale University, all in economics.

WILLIAM N. HUBBARD, JR., recently retired as president of the Upjohn Company. He taught medicine at New York University and served as dean of the University of Michigan Medical School (1959–1970) and professor of internal medicine (1964–1970) before joining the Upjohn Company in 1970. He was elected president of Upjohn in 1974. He is a member of numerous medical honorary societies and has served several medical and educational associations in various capacities. A former member of the National Science Board, he currently serves as a consultant. He is currently chairman of the Council on Health Care Technology of the Institute of Medicine. He received an A.B. degree from Columbia University in 1942 and an M.D. degree in 1944 from New York University.

GLORIA T. JOHNSON has been with the International Union of Electronic, Electrical, Technical, Salaried & Machine Workers, AFL-CIO, since 1954 and is currently director of the Department of Social Action, responsible in particular for education and women's activities. Prior to 1954, she taught for a brief period at Howard University and served as an economist with the U.S. Department of Labor and the Wage Stabilization Board. She is also chair of the IUE Women's Council and in this position is a member of the IUE Executive Board. She serves as chair of the AFL-CIO Committee on Salaried and Professional Women and has been treasurer of the Coalition of Labor Union Women since its founding. She also serves many other educational, labor, women's, and community organizations. She received both B.A. and M.A. degrees from Howard University.

ROBERT E. KRAUT is a social psychologist on the technical staff at Bell Communications Research and an adjunct faculty member in the Department of Psychology at Princeton University. He has previously held positions at Bell Laboratories, Cornell University, and the University of Pennsylvania. His research focuses on the way people judge themselves and others, on interpersonal interaction, and on the social impact of new information technologies. He has a B.A. in English and social relations from Lehigh University and a Ph.D. in psychology from Yale University.

SHIRLEY M. MALCOM is head of the Office of Opportunities in Science of the American Association for the Advancement of Science (AAAS). She served previously as a program officer in the Directorate for Science Education of the National Science Foundation (NSF), in previous positions at AAAS, and as assistant professor of biology at the University of North Carolina, Wilmington. She serves a large number of organizations concerned with educational equity for women, minorities, and disabled persons, science and technology policy, and human resource issues. She has served as chair of the NSF Committee on Equal Opportunities in Science and Technology, as a commissioner of the Commission on Professionals in Science and Technology, and as a member of the Advisory Council of the Carnegie Forum on Education and the Economy and the Carnegie Task Force on Teaching as a Profession. Malcom has a B.S. from the University of Washington and an M.A. from the University of California, Los Angeles, both in zoology, and a Ph.D. in ecology from the Pennsylvania State University.

MICHAEL J. PIORE is professor of economics at the Massachusetts Institute of Technology. He has worked on a variety of problems in labor economics and industrial relations, including technological change, minority workers, migration, and union structure and organization. He developed the dual labor market hypothesis, which is an attempt to explain the economic problems of marginal workers, particularly racial and ethnic minorities, in terms of the types of jobs to which they are confined and the role these jobs perform in the functioning of the economy. More recently he has examined the transformation in economic activity from mass production to flexible specialization. Convinced that the political, social, and historical forces that affect economic institutions are critical, he is currently exploring new forms of business organization, new managerial practices, innovations in union structures and collective bargaining, and changes in the role of government, particularly state and local government, in the economy. Piore holds B.A. and Ph.D. degrees in economics from Harvard University.

FREDERICK A. ROESCH is senior vice-president for Citicorp/Citibank's Institutional Global Electronic Markets activities. He joined Citibank in 1964 and has served as an officer in branches in Japan, Taiwan, and Thailand. In 1973 he was appointed head of Citicorp's worldwide venture capital activities. Named to head personnel planning and development for Citicorp in 1977, he assumed his current position in June 1985. He is a graduate of Dartmouth College and has an M.A. from the University of California, Berkeley.

TERESA A. SULLIVAN is associate professor of sociology at the University of Texas at Austin, where she is also associate director of the Population Research Center and chair of the Women's Studies Steering Committee. Her scholarly work investigates marginal incorporation into the economy and includes work on underemployment, studies of sex discrimination in fringe benefits, and studies of immigrant workers, particularly women. She is currently completing the data collection phase of a large study of consumer bankruptcy. She previously served on the NRC Panel on Immigration Statistics. She has a B.A. from James Madison College of Michigan State University and M.A. and Ph.D. degrees in sociology, both from the University of Chicago.

LOUISE A. TILLY is professor of history and sociology on the Graduate Faculty of the New School for Social Research and chair of its Committee on Historical Studies. Previously she taught at Michigan State University, the University of Michigan, and at the Ecole des Hautes Etudes en Sciences Sociales, Paris. She is member of the Executive Committee of the Board of the Social Science Research Council, the Council of the American Historical Association, and the Panel on Technology and Employment of the Committee on Science, Engineering, and Public Policy at the National Academy of Sciences. Her current research includes a comparative historical study of the state, class, and family in French cities; she is also completing work on the labor force and class formation in late-nineteenth-century Milan. She received an A.B. from Douglass College, an M.A. from Boston University, and a Ph.D. from the University of Toronto.

DONALD J. TREIMAN is professor of sociology at the University of California, Los Angeles. His research interests center on the comparative study of social stratification and social mobility. He has studied problems of occupational classification and measurement extensively, in particular analyzing occupational prestige data from 60 countries. Previously he served as study director of the Committee on Occupational Classification and Analysis at the National Research Council, which produced reports on job evaluation, comparable worth, and the *Dictionary of Occupational Titles;* he was also study director of the Committee on Basic Research in the Behavioral and Social Sciences, which produced two volumes on the value and usefulness of basic research. He has a B.A. from Reed College and M.A. and Ph.D. degrees from the University of Chicago, all in sociology.

ROBERT K. YIN is chairman of the board of COSMOS Corporation, having served as its first president from 1980 to 1985. He conducts research on a variety of topics at COSMOS, a social science research and consulting firm, including the evaluation of neighborhood programs initiated by the Ford Foundation and the MacArthur Foundation; economic development and high-technology firms; and the use of office automation and other advanced technologies (e.g., robotics and artificial intelligence) by organizations. Previously he worked for eight years at the Rand Corporation. He has a B.A. from Harvard University and a Ph.D. in psychology from the Massachusetts Institute of Technology.

PATRICIA ZAVELLA is assistant professor of community studies at the University of California, Santa Cruz. Her research has focused on occupational segregation in the California canning industry, Mexican-American women workers, and the impact of women's production employment in the electronics, apparel, and canning industries on family structure. She is a member of the Silicon Valley Research Group. She has a B.A. degree from Pitzer College and an M.A. from the University of California, Berkeley. Her Ph.D. in anthropology is also from the University of California, Berkeley.

Index

A

Accountants and bookkeepers
  impacts of technological change on, 32, 38–40, 172
  minorities employed as, 92
  number of, 3, 4, 38–40, 91, 94
  projected growth in, 114, 116, 121, 123
  sex stratification of, 20, 40
  unionization of, 38–39
Administrative support occupations
  female employment by race/ethnic origin, 93–95
  number of workers in, 4–5, 67
  sex ratios in, 4–5
  see also Clerical employment/occupations; and specific occupations
AFL-CIO Committee on the Evolution of Work, 17
Agriculture
  clerical staffing ratios, 96–97, 103
  degrees awarded to women in, 78
  displacement of workers, 1, 15, 82
  employment growth in, 102
  occupational groups of women employed in, 18
American Association for Medical Systems and Informatics, 57
American Nurses' Association Council on Computer Applications in Nursing, 57

Asian-Americans, clerical work force, 92–95
Association for the Development of Computer-Based Instructional Systems, 57
AT&T
  EEOC consent decree, 28
  technology-change committees, 29, 163
Automobile manufacturing, productivity and employment in, 15

B

Baby boom cohorts, effect on labor markets, 16, 69, 79
Bank tellers
  career mobility, 46
  employment trends of, 46–48, 88–90, 115–116, 120, 172
  female employment by race/ethnic origin, 95
  number of, 3, 5, 90
Banking
  computerization of, 9, 35, 44–48, 138, 155, 157, 164–165
  educational requirements in, 47–48
  employment growth, 13, 47, 59, 97–100, 106, 118
  female employment in, 13, 45–46
  occupational shifts in, 46, 48

structural change in, 44-48, 106
see also Financial industry
Barriers to employment, 43, 85-86, 169
see also Equal employment opportunity laws; Sex segregation in employment
Bell, Alexander G., 26
Blacks
  in administrative support occupations, 43, 93-95
  educational attainments, 77
  family responsibilities of, 19
  labor force participation rates, 69, 92
  negative effects of automation on, 43
  shifts in employment, 1
  see also Minorities
Bureau of Labor Statistics
  classification of occupations, 66, 138-139
  employment projections, 63, 71, 104-117, 123-125, 143, 180
  female labor force participation rates, 74
Bureau of the Census, classification of occupations, 66, 68, 143

C

Career ladders/mobility
  in banking, 46
  in insurance industry, 42-44
  job segregation effects on, 19, 22
Career opportunities
  expansion of, for women, 173-175
  policies affecting, 64, 66, 68
Cashiers
  growth of employment of, 32, 49-50, 52, 88, 90, 104, 107-108, 111-112, 118, 123
  number of, 3, 90
  quality of jobs for, 135
  reclassification of, 3, 49-50, 107
  sex segregation of, 19-20, 40, 50, 51
Child care, 69-70, 145
Childbearing, 69-70, 72-73
Clerical employment/occupations
  back-office jobs, 22, 41, 119, 126
  declines in, 89, 106
  definition, 2-3
  demographic trends, 89-96
  factors affecting future of, 64, 106
  fastest-growing industries for, 106
  fastest-growing jobs, 118
  in financial industry, 16, 45-46, 63, 92, 96, 103

geographic trends, 92
growth by industry, 96-103
growth, 21, 86-89, 90-91, 96-111, 167
in health industry, 54-58, 99-100
historical patterns in, 32-61
innovations influencing, 32-34; *see also* Innovations/inventions; Word processing
in insurance industry, 13, 16, 40-44, 59, 82, 92, 137
with largest negative employment changes, 119-121
with largest number of workers, 3, 88
with largest projected job growth for women, 121-122
minority representation in, 42-43, 54, 69, 89-90
projections, 21, 47, 70-73, 79-83, 103-126, 174-175
recession effects on, 87, 92, 97, 99, 124
shifts in, 88-89, 111-123, 173; *see also* Occupational shifts
sources of change in, 96-103
staffing ratio changes, 93-103, 106, 110-119
telecommuting, 144-147
see also Administrative support occupations; and specific occupations
Collective bargaining, *see* Unions
Commission of the European Communities, implementation of technology, 160-161
Committee on Science, Engineering, and Public Policy, 81, 171
Computer operators
  employment levels, 3, 4-5, 21, 83, 88, 111-112, 118, 120-121, 123
  minorities employed as, 92-93
Computers/computerization
  in banking, 9, 35, 44-48, 138, 155, 157, 164-165
  in communications industry, 9-10, 27-28
  data entry technologies, 8
  diffusion of, 7-8, 11, 39, 64
  early business applications, 39
  electronic mail, 9, 128
  employment effects, *see* Employment effects of technological change
  in financial industries, 103, 124
  in health industry, 54-58
  in insurance industry, 13, 40-43, 59, 82, 136-137
  output and display technologies, 9

INDEX

projected developments in, 10
projected needs for, 12
resistance to, 8, 55
in retail industry, 49, 51
storage and processing, 8-9
system compatibility and interconnection, 9;
see also Networking
see also Office automation;
Telecommunications;
Telecommuting/telework; Word processing
Current Population Survey (CPS), occupational classifications, 66-67, 89, 124
Customer service representatives
employment rates, 15, 112, 114
in insurance industry, 42
integration of jobs by, 137

D

Data entry operators, employment levels, 95, 117-119
Demographic trends
in clerical employment, 89-96
in labor force, 68-74, 79-81
Department stores, *see* Retail industry
Deskilling
contradictory impressions of, 136-143
of secretarial work, 17, 35, 137
as way of entry for disadvantaged workers, 142
see also Job content; Quality of employment; Skill level changes
Dictionary of Occupational Titles, occupational classifications, 66-67, 138-139
Discrimination, *see* Barriers to employment; Equal employment opportunity laws; Sex segregation in employment

E

Earnings and wages
bookkeepers, 38
compensation for new skills, 131, 149
educational attainment and, 149
effects of technological change on, 83
of home-based workers, 147
job segregation effects on, 19
losses on reemployment, 124-125
productivity and, 149

sex ratio in, 18-19, 80
supermarket workers, 52
Economic conditions and considerations
female employment and, 65, 80, 87
growth and employment, 1, 13-14, 21, 59, 65, 80, 104
quality of employment, 148-157
see also Recessions
Educational attainment
advanced degrees awarded to women, 79
by birth cohort, 76
by fields of study, 77-79, 169
by gender, 77-78
by race/ethnic origin, 77-78
wages and, 149
Educational needs and requirements
in banking, 47
in insurance industry, 44
job transition programs, 176
for minority women, 126, 171, 176
in nursing, 53, 58
for occupations with largest job growth, 123
recommendations of panel regarding, 169-172
to respond to occupational shifts, 81, 169
Educationally disadvantaged
effect of technological change on, 13, 44, 47
programs for, 171
Effective Technical and Human Implementation of Computer-Based Systems (ETHICS), 161
Employment
clerical, *see* Clerical employment/occupations
discrimination in, *see* Barriers to employment; Equal employment opportunity laws; Sex segregation in employment
displacement, *see* Job displacement
government, *see* Government employment, federal and state and local
at home, *see* Home work; Telecommuting/telework
levels, *see* Levels of employment
participation, *see* Participation in labor force
projections, *see* Projections
quality, *see* Quality of employment; Working conditions
rates, *see* Labor force participation rates; Occupational staffing ratios; Projections; and specific industries and occupations

security of, *see* Job security; Job tenure; Layoffs
shifts in, *see* Occupational shifts
of women, *see* Women's employment
see also Unemployment
Employment effects of technological change
computerization, 8, 10, 39, 43, 52, 54, 83, 109, 118, 157, 173–174
demand for workers, 81–88
images of, 13–14, 127–129
measures of, 11, 14–15, 167–168
negative, 11, 16–17, 39, 43, 52, 54, 60, 83–84, 128, 173–176
positive, 11, 17, 37–38, 82, 128
sex differential, 18–23, 32–48, 125–126, 168–169
Equal Employment Opportunity Commission, consent decree with AT&T, 28
Equal employment opportunity laws
effects on women's employment opportunities, 64–65
recommendations of panel on importance of enforcement, 173–175
Equitable Life Assurance Society of the United States, contract for advance notice of automation, 44
Ergonomics, 147–148, 151–153, 155–156, 177–179; *see also* Video display terminals

F

Family responsibilities, 19, 22–23, 145–147, 171
Financial industry
clerical staffing ratios, 96–97, 103
geographic trends in employment, 92
growth in employment, 15–16, 63, 92, 102–103
technological effects on employment, 15, 103, 124
see also Banking

G

Government employment, federal and state and local
clerical declines, 92, 102–103
displacement in, 124
growth in, 15–16, 97, 100, 106
staffing ratios, 96–98, 110
Growth, economic, *see* Economic conditions and considerations; Projections; Recessions

H

Health industry
clerical employment growth in, 98–100, 106
computerization of, 54–58
medical diagnostic systems, 8, 55–58
medical informatics, 55
structural changes and employment growth, 52, 58, 106
see also Nurses/nursing
Hispanics
educational attainments, 77
family responsibilities, 19
labor force participation rates, 69, 92
see also Minorities
Home work, 22, 144–147
Honeywell Corporation, survey of office automation reactions, 132–134
Hunt and Hunt studies, 1, 8, 14, 21, 34, 43, 46, 63, 65, 67, 83, 86–109, 112–118

I

Implementation of technology
adaptation to, 75, 169, 175–177
advance notice of, 31, 44, 160, 162–163, 173
in banking, 35, 47, 155, 164–165
choices in, 12, 59–60, 156–166
clerical workers' influence in, 152
constraints on, 59–60, 148–157, 174, 176
design space concept, 160–161
employee participation programs, 152, 178–179
employment effects of, *see* Employment effects of technological change
government role in, 170
guidelines and planning for, 6, 27–28, 35, 37, 47, 136, 153–154, 164–165, 177–178
in health care industry, 57–58
historical patterns, 24–61
in insurance industry, 136
management dominance and role in, 151–154
negative effects of, 11, 16–17, 19, 22, 28, 39, 43, 52, 54, 60, 83–84, 89, 103, 109, 111, 119, 128, 157, 169, 173–176
by noncomputer professionals, 136
Norwegian Bank Employees Union (BNF) example of, 164–165
positive effects of, 1–2, 6–11, 17, 36–38, 82, 103, 128, 157, 164–165, 169–170

recommendations of panel regarding, 177–179
union involvement in, 153, 162–165
women's participation in, 54–58, 151–153, 158–159, 164–165, 178
worker participation in, 158–162
Innovations/inventions
assembly line, 11
automatic dialing, 27–28
automatic teller machines, 45, 106, 118
bank credit cards, 45
Burroughs adding machine, 32
diffusion of, 7–8, 10–11, 24–27, 39, 64, 109, 170, 174
electronic funds transfer, 45
employment declines from, 83
intelligent cash registers, 49, 51, 111, 135
magnetic ink character recognition, 45
medical monitoring and patient care systems, 8, 55–58
optical character readers, 8, 119
phonograph and radio, 12
punched card and counter-sorter, 32
replacing human interaction, 118
scanning technologies, 8, 49, 135
tabulating machines, 89
telemedicine, 56–57
telephone, 25–29; *see also* Telecommunications; Telephone operators
typesetting equipment, 30–31; *see also* Printing and publishing
typewriter, 32
Xerox photocopy method, 6
see also Computers/computerization; Employment effects of technological change; Telecommunications; Word processing
Insurance industry
automation of claims procedures, 43, 137
career ladders/mobility in, 42, 43–44
computerization of, 13, 40–43, 59, 82, 136–137
employment effects of automation, 40–44, 63, 82
female employment in, 13, 41, 44
growth of employment in, 13, 15–16, 40, 43, 90, 95, 98–100, 112–113, 118, 134
historical patterns of technological change in, 40–44
job integration in, 42

skill-level changes, 137–139, 142
suburbanization of jobs, 92
International Medical Informatics Association, 58

J

Job content
attributes of, 130
fragmentation, 136–139, 149
levels of analysis, 139–140
reorganization and reintegration, 42, 137, 167
routinization, 138–139
stages of technology, 139–140
variety in, 130, 140
see also Deskilling; Job satisfaction; Quality of employment; Skill-level changes
Job displacement
of agricultural workers, 15
in clerical work, 62, 124–125
due to shifts in demand, 84, 172
due to technological advances, 59, 172
geographic mobility and, 86, 167
see also Earnings and wages; Employment effects of technological change; Location of work; Unemployment
Job integration in insurance industry, 42
Job satisfaction
influences on, 150
office automation effects on, 129, 132–136, 168
surveys of, 132–136, 149, 168
worker implementation of technology and, 136, 158
Job security
advance notice of implementation of new technology, 31, 44, 160, 162–163, 173
agreements, 31, 44
effects of technology on, 131, 135, 142
recommendations of panel regarding, 172–173
ways for employers to provide, 149–150
Job segregation, *see* Sex segregation in employment
Job tenure, 75; *see also* Job security; Work-life expectancy

K

Kelly Services, survey of office automation reactions, 132–133, 149

L

Labor force growth, projections, and size, *see* Participation in labor force; Projections
Labor force participation rates
  accuracy, 70
  by age, 72–74
  by gender and race, 71–72
  in male-dominated occupations, 125
  projections, 68–72, 174
  see also Participation in labor force
Layoffs, advance notice of, 29; *see also* Job security
Leontief-Duchin studies, 1, 62, 108–111
Levels of employment
  productivity growth and, 15–17
  sources of change in, 16, 25
  see also Clerical employment/occupations; Employment effects of technological change; Participation in labor force; Occupational shifts; Projections; Unemployment
Location of work
  constraints on women, 22, 44, 86
  effects of technology on, 11
  geographic trends, 92
  see also Job displacement; Unemployment, structural

M

Management/managers
  constraints on, in implementation of technology, 154–157
  effects of technological change on, 150
  employment growth for, 109
  female percentage, 152
  reintegration of jobs of, 137–138
  role in implementation of technology, 37, 151–152
Minolta Corporation/Professional Secretaries International, survey of office automation reactions, 132–134
Minorities
  education and, 77–78
  negative effects of technology on, 16–17, 19, 22, 54, 169
  representation in clerical occupations, 42–43, 54, 91–92
  see also Asian-Americans; Blacks; Hispanics
Monitoring employee performance, 17, 27, 44, 128, 131, 143–144

N

National League on Nursing, 57–58
National Research Council
  Board on Telecommunications and Computer Applications, 145
  Committee on the Effective Implementation of Advanced Manufacturing Technologies, 153
  Committee on National Statistics, 14
  Committee on Vision, 130, 147–148, 179
Networking, 9, 64, 174
9-to-5, survey of job stress among women, 132–134, 144, 151
Nurses/nursing
  educational programs and trends, 53, 55
  effects of technology on, 54–58
  implementation of technology, 54–58, 158–159
  minority employment, 54
  numbers of, 52–53, 123
  sex differentials in, 53–54
  stress in, 55–56
  unemployment rates, 53, 58

O

Occupational Employment Statistics (OES), 67, 111
Occupational shifts
  in banking, 46
  in clerical work, 88–89
  education, training, and retraining for, 170–172
  in health industry, 58
  in insurance industry, 42
  from manufacturing to service, 15–16
  projections for, 111–123, 173
  in retail industry, 49–50
  in telecommunications industry, 28
Occupational staffing ratios
  clerical variation across industries, 96–103
  negative effects of changes in, 118–119
  ranking of clerical occupations by, 111–117
  relative effects of, 106, 110–111
Occupations
  classification of, 60–61, 66, 68, 107, 109–110, 138–139, 142–143

with largest job growth, 122-123, 142
reallocation of functions among, 82; *see also* Occupational shifts
sources of change in size of, 96
see also Clerical employment/occupations, and specific occupations
Office automation
impact on clerical employment, 21, 63, 107, 118, 137-138
job satisfaction and, 129
productivity and, 36-37, 128
see also Computers/computerization; Employment effects of technological change; Ergonomics; Implementation of technology; Innovations/inventions; Quality of employment; Word processing
On-the-job training, sex segregation in, 22, 46, 171
Output, *see* Productivity

P

Part-time employment, increases in, 50-51
Participation in labor force
by age, 68, 72-73, 80
demand for workers, 81-83
educational attainment and, 75-79
factors affecting, 69-70, 79-81, 86
positive influences on, 69-70
by race/ethnic origin, 69
rates, 68-70; *see also* Labor force participation rates
substitution for women workers, 80
supply of women workers, 68-79, 81, 125
see also Clerical employment/occupations; Telecommuting/telework
Policy recommendations
adaptive job transitions, 177
data and research needs, 181
design, implementation, and application of technology, 177-178
education, training, and retraining, 172
employment security and flexibility, 173
expansion of job opportunities, 175
health concerns, 179
worker participation in implementation of technology, 178-179
Postal clerks
employment levels, 21, 91, 94, 103, 117-118
minorities employed as, 92

Printing and publishing industries
clerical employment in, 91, 95, 98, 113-114
female work force, 30-31
historical patterns of technological change in, 29-32
home-based workers, 145
innovations in, 30-31
job segregation in, 29-32
Productivity
employment levels and, 15-17, 81-82
gains from new technology, 64, 82-83, 109, 156-157, 174, 176
growth in, 13
innovations increasing, 14, 64
measures of, 14, 143-144
monitoring, 143-144
office automation and, 36-37, 82-83, 123
wages and, 149
Project on Connecticut Workers and Technological Change, 128
Projections
accuracy of, 70
Bureau of Labor Statistics, 63, 71, 104-117, 123-125, 143, 180
data availability for, 66-68, 179-180
demand for workers, 81-83, 174-175
Hunt and Hunt, 103-109, 112-118
labor force participation rates, 70-73
labor supply, 79-81, 174
largest projected clerical job growth for women, 121-122
Leontief-Duchin, 1, 62, 108-111
levels of clerical employment, 21
occupational shifts in clerical work, 111-123
overall employment growth, 103-111
of panel, 167-170, 172-175
problems in, 63-65, 69, 109-110
unemployment, 83-86
women's wages, 18-19
see also specific industries and occupations

Q

Quality of employment
assessment of, 67-68, 131-136
biases in assessing, 132-136
changes in, *see* Deskilling; Skill-level changes
concerns about, 17-18, 127-128

criteria for assessing, 129–131
definition of, 129–131
economic considerations in, 131, 148–150, 154, 156–157; *see also* Job security
employer benefits in ensuring, 154–155
equipment age and, 140
improvements in, 17, 44, 60
management's role in enhancement of, 153–154
occupational mix changes, 129
recommendations of panel for monitoring health concerns, 179
sources of information on, 131–136
word processing effects on, 34, 140–141; *see also* Office automation
see also Job content; Job satisfaction; Working conditions
Quality of Employment Survey, 67, 131, 180

R

Recessions, employment effects of, 87, 92, 97, 99, 124
Recommendations, *see* Policy recommendations
Retail industry
career opportunities, 51
cash register innovations and computerization, 49, 51
clerical staffing ratios, 96–97
deskilling in, 52, 137
employment growth, 15, 50, 123
food sector trends, 49, 51–52
growth in clerical employment, 96–98, 101–102
occupational shifts, 49–50, 52
structural changes in, 48–49, 51–52
turnover rates, 51
unionization, 51–52

S

Sales workers, *see* Cashiers
Seasonal employment, 51
Secretaries
blurring of lines between managers and, 11, 35
deskilling of, 17, 35, 137, 142
diverse duties of, 33
growth rates for, 90, 115, 118, 120–121

number of, 3, 4, 33–34, 37–38, 67, 90, 93
replacement of by paraprofessionals, 35
word processing impacts on, 11, 34–37; *see also* Office automation
see also Clerical employment/occupations
Service sector employment
clerical staffing ratios, 96
employment levels, 97
growth in, 13, 15–16, 21, 102, 104
see also specific industries and occupations
Sex segregation in employment
career mobility and, 19
in data processing, 40
extent of, 19–20, 80
in on-the-job training, 22, 46
in printing and publishing, 29–31
in retail industry, 50
in wages, 18–19, 80
see also Barriers to employment; Equal employment opportunity laws
Sex stereotypes, 22, 23
Shifts in employment, *see* Occupational shifts
Skill-level changes and mismatches, 17, 60, 82–83, 136–139, 141, 168; *see also* Deskilling
Social Security Administration, implementation of technology by, 157
Stenographers, employment levels, 4, 21, 83, 89, 91, 93, 118, 120
Swedish Act of Codetermination, 160

T

Tabulating machine operators, 89
Teachers' aides
growth in, 88, 90, 112, 120–121
minority employment as, 92, 95
Telecommunications
automatic dialing, 27–28
automatic switching, 10, 27–28, 103
clerical employment in, 98; *see also* Telephone operators
effects on information processing jobs, 10
equipment innovations, 9–10, 24–27
historical patterns of technological change in, 25–29
occupational shifts in, 28, 89
productivity and employment in, 15
projected developments in, 10, 29
Telecommuting/telework, 144–147

INDEX *215*

Telephone operators
  employment trends, 15, 27-28, 89, 91, 94, 103, 111, 117-118, 120
  monitoring performance of, 17, 27
Temporary help industry, growth projections, 122
Training needs, 13, 172; *see also* Educational needs and requirements; On-the-job training
Turnover rates
  department store industry, 51
  women workers, 23, 27, 60-61, 171
Typists
  growth rates, 88, 90, 116, 120
  minorities employed as, 92-93
  number of, 3, 4, 21, 90
  see also Office automation

U

Unemployment
  cyclical increases in, 16, 84-87, 109
  projections, 83-86
  public policy impacts on, 85
  rates for nurses, 53
  solutions to, 85, 175-177
  structural, 84-85, 111, 125
  transitional, 175
  types, 83-84
  see also Job displacement
Unions
  Communications Workers of America, 29
  International Typographical Union, 30-31
  negotiated technology agreements, 162
  Norwegian Bank Employees Union, 164-165
  Office and Professional Workers, 38
  retail industry activities of, 51-52
  role in adoption of new technology, 153, 160, 162-165
  Service Employees International Union, 44
  State, County, and Municipal Workers of America, 39
  technology change committees, 29, 163
  United Federal Workers, 38
  United Food and Commercial Workers, 51-52
  women's access to, 22
  Women's Trade Union League, 27

U.S. National Commission on Technology, Automation, and Economic Progress, 16-17, 127

V

Video display terminals (VDTs), 17, 44, 68, 128, 134-135, 154, 156, 179; *see also* Ergonomics

W

Wages
  two-tier, collective bargaining for, 52
  see also Earnings and wages
Wholesale trade, employment in, 97, 100, 102, 110, 123, 147-148
Women's employment
  clerical occupations, *see* Clerical employment/occupations, and specific occupations
  differential effects of technology on, 18-23, 32-48, 125-126, 168-169
  earnings and wages, 18-19, 38, 80, 83
  economic considerations in, 65, 80, 87
  educational attainment and, 75-79, 169-170, 172
  expansion of opportunities for, 64-65, 173-175; *see also* Barriers to employment
  family responsibilities and, 19, 22-23, 145-147, 171
  in financial industry, 40-44
  in insurance industry, 52-58
  location of work and, 22, 44, 86
  in nursing, *see* Nurses/nursing
  occupations with largest projected growth, 121-122
  on-the-job training, 22, 46, 171
  in printing and publishing, 29-32
  by race/ethnic origin, 93-95
  in retail trade, 50
  sex segregation in, 19-20, 22, 29-31, 46, 80
  stress in, 132-134, 144, 151
  supply of workers, 68-79; *see also* Labor force participation rates
  in telecommunications, 25-29
  turnover rates, 23, 27, 60-61, 171
  in unions, *see* Unions
  see also Participation in labor force; Projections; and specific occupations

Word processing
   assessing use of, 66–67
   classification of, 66
   employment increases due to, 11
   fragmentation of jobs by, 137
   impact on secretaries, 11, 34–37
   productivity impacts of, 36, 82, 109
   social organization of, 139–140
   variety of tasks in, 140–141
   see also Office automation
Worker satisfaction, *see* Job satisfaction
Working conditions
   health concerns, 134–135, 144, 147–148, 168, 179
   home-based clerical work, 145–147
   physical, 130, 147–148, 151–152, 154
   social, 130–131, 155
   stress, 17, 27, 55–56, 132–135, 144
   see also Ergonomics; Monitoring employee performance; Quality of employment; Video display terminals
Working Women Education Fund, 17
Work-life expectancy, 75